LECTURES ON QUANTUM THEORY

LECTURES ON QUANTUM THEORY

Mathematical and Structural Foundations

Chris J. Isham
Imperial College, London

ICP
Imperial College Press

Published by

Imperial College Press
57 Shelton Street
Covent Garden
London WC2H 9HE

Distributed by

World Scientific Publishing Co. Pte. Ltd.
5 Toh Tuck Link, Singapore 596224
USA office: 27 Warren Street, Suite 401-402, Hackensack, NJ 07601
UK office: 57 Shelton Street, Covent Garden, London WC2H 9HE

Library of Congress Cataloging-in-Publication Data
Isham, C. J.
Lectures on quantum theory : mathematical and structural foundations / Chris J. Isham.
p. cm.
Includes bibliographical references and index.
ISBN 1860940005 1860940013 (pbk.)
1. Quantum theory--Mathematics. 2. Mathematical physics. I. Title.

QC174.17.M35 I84 1995 95025184
530.1'2--dc20

British Library Cataloguing-in-Publication Data
A catalogue record for this book is available from the British Library.

First published 1995
Reprinted with a larger typeface 1997, 2001, 2004

Printed in Singapore by Utopia Press Pte Ltd

Preface

This book is based on notes handed out to students at Imperial College who attended my undergraduate lecture course on quantum theory during the years 1989 to 1994. All students in the Physics Department take at least two courses in quantum theory: one in their first year, and one in their second. My course was part of the third year theoretical physics option. Thus, in preparing the notes, I could assume a genuine interest in mathematical approaches to physics, plus a moderate knowledge of basic wave mechanics, and a small exposure to the ideas of electron spin and the Pauli spin matrices.

The aim of my lectures was to build on this background to provide an introduction to the structural foundations of modern quantum theory. The scope is deliberately limited to the non-relativistic regime of physics, but the text includes discussions of many of the basic conceptual problems that arise in this context. The main mathematical tool is the theory of operators on vector spaces, and this is developed from scratch, albeit without becoming too entangled in mathematical detail. For example, although I mention infinite-dimensional spaces, no effort is made to describe the necessary analysis in any rigorous way. This is not because I doubt the importance of these matters, but too many mathematical subtleties can distract attention from the fundamental issues.

On the other hand, these notes are not intended to be a textbook on the philosophical foundations of quantum theory. Several comprehensive works of this type have appeared recently, and I am not trying to rival them. Nor do I have any personal axe to grind concerning interpretational issues in quantum theory (other, perhaps, than prejudices stemming from my interest in quantum gravity): the course was intended to provide an introduction to some of the issues in a reasonably balanced, but brief, way. Indeed, although a number of explanatory paragraphs have been added to the material from which I spoke, I have deliberately kept to the abbreviated style of the original lecture notes: the course was only twenty-five lectures long (I do talk rather quickly!), and this is reflected in the length of the book. I have included the problems and worked answers that were handed out to the students to assist them in mastering the basic mathematical tools of vector-space theory. This is an important part of the overall development of the work and a self-study student should benefit from looking at them seriously.

I would like to express my thanks to all those students who attended my course and drew my attention to various errors and confusions in the earlier versions of the notes. I am particularly grateful to Steve Weinstein for his extensive and penetrating remarks on the conceptual aspects of quantum theory and the way they are treated here.

Finally, I would like to dedicate this work with great affection to my wife Valerie, and to my daughters Nicola and Louise. For many years they have had to cope with the stress of being married to, respectively fathered by, a theoretical physicist. Their collective survival skills are most impressive.

Chris Isham

Imperial College, March 1995

Contents

Chapter 1

INTRODUCTION

1.1 Scope of the Book

The use of mathematics in modern physics can be rather daunting. In classical mechanics, the representation of the position of a particle by a euclidean three-vector can be directly related to our immediate impressions of the world around us. However, quantum theory is distinguished by a striking gap between the mathematical structure and the physical objects it seeks to represent: a gap that can become an unbridgeable chasm for students encountering the subject for the first few times. For this reason, it is worth being clear in advance what the main goals are in the material that follows.

A theoretical structure in modern physics typically has the following, quasi-axiomatic, features:

1. a specified domain of applicability that limits the class of physical systems to which the theory should be applied;

2. an identification of certain physical concepts that relate to the class of systems in this domain;

3. a specification of the general mathematical framework within which the theory is to be presented;

4. a collection of rules that relate the physical concepts to elements of the mathematical structure;

5. an overall conceptual scheme for analysing the meaning of fundamental terms employed in the statement of the rules;

6. a set of techniques for applying the rules to specific physical systems within the class admitted by the domain of applicability.

There are various criteria by which a structure of this type is judged. For example,

1. the domain of applicability should be as large as possible;

2. the theory should be empirically effective within that domain;

3. the techniques used in formulating the theory (some of which may involve new mathematics) should be internally consistent in a mathematical sense;

4. the physical and conceptual ideas used in elucidating the 'meaning' of the theory should be consistent in a genuine philosophical sense;

5. the overall formalism should be simple and elegant.

The original domain of applicability of quantum theory was the area of non-relativistic atomic and molecular physics, particularly systems of a finite number of elementary particles moving under the influence of electrostatic binding forces. The scope of quantum theory has been slowly developed since then to include relativistic atomic and subatomic processes, mainly within the context of quantum field theories. Schemes of this type have been very successful empirically, but certain problems arise (the 'ultra-violet' divergences) that still cause disquiet at a fundamental level. Attempts to extend quantum theory to include *general* relativity have encountered major mathematical and conceptual problems, and it is not yet clear what the outcome will be. In this book, the domain of applicability is limited to non-relativistic physics, although many of the ideas can be extended to include the relativistic case.

Many approaches to quantum theory take on a rather operationalist flavour by focussing on the concepts of 'observable' and 'measurement'. The emphasis is then placed on the instrinsically probabilistic nature of the predictions of the results of such measurements. This approach is also adopted in the first half of this book; later on we shall look at some aspects of what happens in less instrumentalist interpretations of the theory. In either case, the appropriate mathematical tools for quantum theory[1] are based on the theory of Hilbert spaces: finite- or infinite-dimensional vector spaces equipped with an inner product that mimics the overlap function of elementary wave mechanics.

The book begins with a short summary of some of the key ideas in wave mechanics, presented in a way that is adapted to the later discussion of the general quantum formalism. Chapters 2 and 3 then introduce the basic ideas of vector spaces and linear operators needed in the vector space approach to quantum theory. This material is kept to the minimum necessary for the task in hand: in particular, no attempt is made to deal rigorously with the mathematical problems that arise when infinite-dimensional spaces are treated properly.[2]

One of the main goals of the book is to explore some of the deep conceptual issues that arise in quantum theory. Of the many ways in which this topic can be approached, I have chosen to focus on the notion of a *physical property* and the extent to which it is, or is not, meaningful to talk about a quantum system 'possessing' such properties. For this reason, the exposition of quantum theory proper is preceded by a discussion in Chapter 4 of the analogous situation in classical physics: in particular, the way in which a certain philosophical view about the physical world is reflected in the mathematical framework used to describe it. In classical physics, the 'realist' and 'instrumentalist' views of science fit together seamlessly, whereas in quantum physics they differ sharply, especially in their attitudes towards the idea of physical properties. That such a distinction can arise at all is closely tied to the different mathematical structures employed in the formulations of classical and quantum physics.

The section on classical ideology is followed by two chapters that discuss in detail the general rules of quantum theory as they appear in the vector space formalism. The material covered includes vector states and mixed states, compatible operators, time development, unitary operators,

[1]The classic reference is von Neumann (1971).

[2]A more detailed exposition of the type of vector space theory used in quantum theory can be found in my book Isham (1989). A much more comprehensive treatment of the mathematics of quantum theory is contained in the excellent work by Reed & Simon (1972).

the derivation of the canonical commutation relations, and the generalised uncertainty relations. The main emphasis in these sections is on the basic mathematical framework, and therefore I have adopted a pragmatic, rather instrumentalist view of quantum theory that leaves open the question of what other types of interpretation are possible.

The interpretation of quantum theory is addressed in Chapter 8 where we turn to conceptual matters in a more formal way. The material is organised around four major topics: the meaning of probability, the role of measurement, reduction of the state vector, and quantum entanglement. These issues are of fundamental importance in any attempt to find a more realist interpretation of quantum theory: a key issue for anyone who, like myself, is interested in quantum cosmology. This challenge of realism is studied directly in the final chapter that deals with the status of 'properties' in quantum theory. The main topics discussed are the Kochen–Specker theorem, a short introduction to quantum logic, and the Bell inequalities.

The book concludes with a number of worked problems aimed at developing facility in the type of mathematical manipulations that are essential for any theoretical physicist who wants to use the vector space approach to quantum theory. Several worked problems are also included in the text proper as an aid to understanding various pieces of general formalism.

A quick note on references. The rather short bibliography reflects the origin of the book as lecture notes for an undergraduate course. For this reason, I have concentrated on citing papers and books that should be accessible to an advanced physics undergraduate and which, if dipped into, will genuinely enhance his or her understanding without needing a lifetime of study devoted to the task. As emphasised in the Preface, this is intended to be a short textbook for undergraduates—it is not meant to be a definitive review of modern quantum theory!

1.2 A Summary of Wave Mechanics

It is useful to begin by summarising some of the basic formalism of elementary wave mechanics. This will be presented, with the minimum of comment, in the form of four rules that will be generalised in Chapter 5 to apply to arbitrary quantum systems. For simplicity, I shall present the ideas for motion in one dimension; the extension to three dimensions is straightforward and involves no new principles. In writing these rules we recall the following:

- A number a is an *eigenvalue* of a differential operator $\hat{A}$ if it satisfies the differential equation

$$\hat{A}u(x) = au(x) \tag{1.1}$$

  plus appropriate boundary conditions on the function $u(x)$ (for example, that it be square-integrable[3]). The function $u(x)$ is said to be an *eigenfunction* of $\hat{A}$ associated with the eigenvalue a.

- A *self-adjoint* (or *hermitian*) operator $\hat{A}$ is one for which[4]

$$\int_{-\infty}^{\infty} (\hat{A}\psi)^*(x)\phi(x)\,dx = \int_{-\infty}^{\infty} \psi^*(x)(\hat{A}\phi)(x)\,dx \tag{1.2}$$

  for all square-integrable wave functions ψ and ϕ.

With this in mind, the four rules are as follows.

Rule 1. The quantum state of a point particle moving in one dimension is represented by a complex-valued wave function $\psi(x)$ that can be normalised to one:

$$\int_{-\infty}^{\infty} |\psi(x)|^2\,dx = 1. \tag{1.3}$$

A crucial idea in quantum theory is that any pair of wave functions ψ_1 and ψ_2 can be *superimposed* with arbitrary complex coefficients α_1 and α_2 to give a new wave function $\alpha_1\psi_1(x) + \alpha_2\psi_2(x)$ (provided that α_1 and α_2 are chosen such that this new function is normalised to one).

Rule 2. Any physical quantity that can be measured (*i.e.*, an observable) is represented by a linear differential operator that acts on the wave functions, and is self-adjoint.

[3]A function ψ is *square-integrable* if $\int_{-\infty}^{\infty} |\psi(x)|^2\,dx < \infty$.

[4]The symbol * denotes the complex conjugate.

Rule 3. (i) The only possible result of measuring an observable A is one of the eigenvalues of the self-adjoint operator $\widehat{A}$ that represents it.

(ii) Assume for simplicity that there is no degeneracy (*i.e.*, any two eigenfunctions of $\widehat{A}$ with the same eigenvalue are proportional to each other) and that $\widehat{A}$ has only a discrete set of eigenvalues $a_1, a_2, \ldots$, with corresponding eigenfunctions $u_1, u_2, \ldots$. Then, if the quantum state is $\psi(x)$, the probability that a measurement of A will yield a particular eigenvalue a_n is

$$\boxed{\text{Prob}(A = a_n; \psi) = |\psi_n|^2} \tag{1.4}$$

where the complex numbers ψ_n are the coefficients in the expansion

$$\psi(x) = \sum_{n=1}^{\infty} \psi_n u_n(x) \tag{1.5}$$

of the wave function $\psi(x)$ as a linear combination of the (normalised) eigenfunctions of $\widehat{A}$.

Rule 4. The state function evolves in time according to the time-dependent Schrödinger equation

$$\boxed{i\hbar \frac{\partial \psi(x,t)}{\partial t} = \widehat{H}\psi(x,t)} \tag{1.6}$$

where the Hamiltonian operator $\widehat{H}$ is obtained from the classical energy expression

$$H = \frac{p^2}{2m} + V(x) \tag{1.7}$$

by replacing the momentum p and position x by their corresponding operators $\widehat{p}$ and $\widehat{x}$.

Comments

1. The eigenvalues and eigenfunctions of a self-adjoint operator have three crucial mathematical properties that are central to their use in quantum theory:

- The eigenvalues are *real* numbers (as they should be if they are to correspond to the results of measurements).
- The eigenfunctions form a *complete* set, *i.e.*, any wave function can be expanded uniquely as in Eq. (1.5).
- The normalised eigenfunctions satisfy the *orthogonality* condition

$$\int_{-\infty}^{\infty} u_m^*(x) u_n(x)\, dx = \delta_{mn} \tag{1.8}$$

where δ_{mn} is the Kronecker delta, defined to equal 1 if $m = n$, and 0 otherwise.

One consequence is that the expansion coefficients ψ_n in Eq. (1.5) can be calculated explicitly from the wave function $\psi(x)$ as

$$\psi_n = \int_{-\infty}^{\infty} u_n^*(x) \psi(x)\, dx. \tag{1.9}$$

In particular, this shows that the expansion coefficients are *unique*.

It can be shown from these results that, for any pair of wave functions ψ and ϕ,

$$\int_{-\infty}^{\infty} \psi^*(x) \phi(x)\, dx = \sum_{n=1}^{\infty} \psi_n^* \phi_n \tag{1.10}$$

and hence, in particular,

$$\int_{-\infty}^{\infty} |\psi(x)|^2\, dx = \sum_{n=1}^{\infty} |\psi_n|^2. \tag{1.11}$$

Thus, if ψ is normalised as in Eq. (1.3), we see that

$$\boxed{\sum_{n=1}^{\infty} \text{Prob}(A = a_n; \psi) = 1} \tag{1.12}$$

which is necessary for the probability interpretation to be consistent (the probability that *some* result is obtained for any measurement must be one).

2. The basic interpretative Rule 3 implies that the long term average value of the results of repeated measurements of an observable A is

$$\langle A\rangle_\psi = \int_{-\infty}^{\infty} \psi^*(x)(\hat{A}\psi)(x)\,dx. \tag{1.13}$$

3. As they stand, the rules above are incomplete since they give no information on how to construct the actual operator that represents any specific observable for a given physical system. This is usually done by invoking the 'substitution rule' which says that the operator that represents the classical observable $A(x,p)$ is $A(\widehat{x},\widehat{p})$ (an ambiguous expression, as we shall see later). This has been invoked already in Rule 4 by requiring the quantum Hamiltonian operator to be given by operator substitution in Eq. (1.7).

However, the theoretical structure is still incomplete since no specification has been given of the operators $\widehat{x}$ and $\widehat{p}$. This is part of the far more general question of what it means to construct the 'quantum analogue' of any given classical system. In the context of elementary wave mechanics, this gap is filled by postulating the operators that represent position and momentum to be

$$(\widehat{x}\psi)(x) = x\psi(x) \tag{1.14}$$

$$(\widehat{p}\psi)(x) = -i\hbar\frac{d\psi}{dx}(x) \tag{1.15}$$

which satisfy the famous 'canonical commutation relation'

$$[\,\widehat{x},\widehat{p}\,] = i\hbar. \tag{1.16}$$

4. Angular momentum plays an important role in many different areas of quantum theory, and is treated in wave mechanics using the operators defined above. Specifically, the components of angular momentum in classical physics are $L_x = yp_z - zp_y$, $L_y = zp_x - xp_z$ and $L_z = xp_y - yp_x$, and it is assumed that the corresponding quantum operators are formed using the substitution rule, so that $\widehat{L}_x = \widehat{y}\widehat{p}_z - \widehat{z}\widehat{p}_y$, $\widehat{L}_y = \widehat{z}\widehat{p}_x - \widehat{x}\widehat{p}_z$ and $\widehat{L}_z = \widehat{x}\widehat{p}_y - \widehat{y}\widehat{p}_x$.

It is straightforward to show from Eq. (1.16) that the angular momentum operators have the commutation relations

$$[\,\widehat{L}_x,\widehat{L}_y\,] = i\hbar\widehat{L}_z \tag{1.17}$$

$$[\,\widehat{L}_y, \widehat{L}_z\,] = i\hbar\widehat{L}_x \tag{1.18}$$
$$[\,\widehat{L}_z, \widehat{L}_x\,] = i\hbar\widehat{L}_y \tag{1.19}$$

which, in turn, imply that

$$[\,\widehat{L}\cdot\widehat{L}, \widehat{L}_x\,] = [\,\widehat{L}\cdot\widehat{L}, \widehat{L}_y\,] = [\,\widehat{L}\cdot\widehat{L}, \widehat{L}_z\,] = 0, \tag{1.20}$$

where the total angular-momentum operator $\widehat{L}\cdot\widehat{L}$ is defined as the sum $\widehat{L}_x\widehat{L}_x + \widehat{L}_y\widehat{L}_y + \widehat{L}_z\widehat{L}_z$. The vanishing commutators of Eq. (1.20) mean that simultaneous eigenfunctions exist of $\widehat{L}\cdot\widehat{L}$ and any one[5] of $\widehat{L}_x$, $\widehat{L}_y$ or $\widehat{L}_z$. For various psychological reasons, the usual choice is $\widehat{L}_z$, with the z-axis being drawn pointing upwards.

One of the less gripping tasks in a course on wave-mechanics is to compute the explicit form of these simultaneous eigenfunctions, and to find their associated eigenvalues. The well-known result is that the eigenvalues of $\widehat{L}\cdot\widehat{L}$ have the form $\ell(\ell+1)\hbar^2$, where ℓ is an integer with $\ell \geq 0$. For any given value of ℓ, the associated eigenvalues of $\widehat{L}_z$ are of the form $m\hbar$, where the integer m ranges from $-\ell$ to $+\ell$ in steps of 1. The corresponding simultaneous eigenfunctions u_{lm} are the famous *associated Legendre polynomials*.

1.3 Beyond Introductory Wave Mechanics

The rules and postulates of wave mechanics have been used widely, and with considerable empirical success. However, a number of subtle and important issues wait to be uncovered. For example:

- What is the precise meaning of 'probability' as it arises in the context of quantum theory? And why does 'measurement' play such a prominent role? Can a measurement be regarded as just another type of physical interaction, or does it need to be considered as a fundamental concept in the very foundations of the theory? If the latter is true, how can this be reconciled with the fact that real measuring devices are composed of atoms, which certainly need to be described in quantum-mechanical terms?

[5]It is only *one* of the operators $\widehat{L}_x$, $\widehat{L}_y$ or $\widehat{L}_z$ since a simultaneous eigenfunction of, for example, the pair of operators $\widehat{L}\cdot\widehat{L}$ and $\widehat{L}_z$ will generally *not* be a simultaneous eigenfunction of the pair $\widehat{L}\cdot\widehat{L}$ and $\widehat{L}_x$, or the pair $\widehat{L}\cdot\widehat{L}$ and $\widehat{L}_y$.

- Do the representations of $\widehat{x}$ and $\widehat{p}$ in Eq. (1.14–1.15) (and the commutation relation in Eq. (1.16)) need to be postulated, or can they be *derived* in some way from the general framework of the theory?

- Are there any other pairs of operators $\widehat{x}$ and $\widehat{p}$ that satisfy the commutation relation Eq. (1.16). If so, do the physical predictions depend on the choice made? In other words, what is more important: the general form of the commutation relation Eq. (1.16), or the specific representations Eq. (1.14–1.15) of the basic observables? Other operators satisfying Eq. (1.16) certainly exist. For example, there is an alternative version of wave mechanics in which states are represented by functions ϕ of momentum p (rather than position x), and the basic operators $\widehat{x}$ and $\widehat{p}$ are defined by

$$(\widehat{x}\phi)(p) = i\hbar\frac{d\phi}{dp}(p) \tag{1.21}$$

$$(\widehat{p}\phi)(p) = p\phi(p) \tag{1.22}$$

which, like the operators defined in Eq. (1.14–1.15), satisfy the canonical commutation relation Eq. (1.16). It is usually assumed that both forms of wave mechanics give the same physical answers. But why should this be so?

- What is the meaning of the 'uncertainty relation' $\triangle x \triangle p \geq \frac{1}{2}\hbar$ associated with the commutation relation Eq. (1.16)? Does $\triangle x$ refer to 'the unavoidable disturbance in the act of making a measurement', as is often stated, or is there some other way of understanding this relation?

- Can the uncertainty relation be generalised to apply to any pair of operators whose commutator is non-zero? Examples are the angular momentum operators $\widehat{L}_x$, $\widehat{L}_y$ and $\widehat{L}_z$ that satisfy the cyclic commutation relations in Eq. (1.17–1.19).

Questions of this type open a Pandora's box of problems concerning the interpretation of quantum theory and the picture it gives of physical reality. Any serious attempt to tackle these problems inevitably encounters a number of profound philosophical issues that are still the subject of intense debate and controversy. One of the central goals of this course is to provide an introduction to some of these deep features of the quantum view of the world.

However, even at the technical level, the postulates above are deficient in several ways, not the least of which is that they apply only to a limited class of physical systems. It is straightforward to extend the wave-mechanical formalism to a particle moving in three dimensions, when the state is a function $\psi(\mathbf{x})$ of the particle's position vector $\mathbf{x}$, or even to a collection of N-particles moving in three dimensions, in which case the state is a function $\psi(\mathbf{x_1}, \mathbf{x_2}, \ldots, \mathbf{x_N})$ of the N position vectors $\mathbf{x_1}, \mathbf{x_2}, \ldots, \mathbf{x_N}$. But there are many important physical systems whose quantum states cannot be described at all using only wave functions.

One example is relativistic quantum physics in which the number of particles can change as a result of interactions between them. For example, consider a scattering experiment in which two particles collide and turn into three particles. Ignoring internal and spin quantum numbers, the initial and final states could be described by wave functions $\psi(\mathbf{x_1}, \mathbf{x_2})$ and $\phi(\mathbf{y_1}, \mathbf{y_2}, \mathbf{y_3})$ respectively. However, it is by no means obvious what type of time-dependent Schrödinger equation could allow a function of two variables to evolve smoothly into a function of three variables.

Another famous example of a system that cannot be described using wave functions is electron spin. If one concentrates purely on its internal-spin properties, a state of an electron can be described by a column matrix[6] $\binom{a_1}{a_2}$, where the analogue of the normalisation condition Eq. (1.3) is that the complex numbers a_1 and a_2 satisfy $|a_1|^2 + |a_2|^2 = 1$. A pair of such states $\binom{a_1}{a_2}$ and $\binom{b_1}{b_2}$ can be superimposed with arbitrary complex coefficients α and β to give a new state $\alpha\binom{a_1}{a_2} + \beta\binom{b_1}{b_2}$ which is defined[7] to be the column matrix $\binom{\alpha a_1+\beta b_1}{\alpha a_2+\beta b_2}$ (the only restriction on α and β is that the normalisation condition must be preserved).

In this system, observables are represented by 2×2 complex hermitian[8] matrices that act on the states by matrix multiplication: an operation that is *linear* with respect to the superposition rule mentioned above. In

[6]A column matrix is often called a column *vector*, reflecting the fact that, as we shall see later, it can be thought of as an element of a particular vector space.

[7]The significance of this definition will emerge in the next chapter.

[8]A square matrix A is said to be *hermitian* if $A = (A^T)^*$ where A^T denotes the transpose of A. In terms of matrix elements, $A_{ij} = A^*_{ji}$.

particular, the x, y and z components of the spin angular momentum are represented respectively by the matrices $S_x = \frac{\hbar}{2}\sigma_x$, $S_y = \frac{\hbar}{2}\sigma_y$ and $S_z = \frac{\hbar}{2}\sigma_z$, where σ_x, σ_y and σ_z are the well-known Pauli spin matrices:

$$\sigma_x = \begin{pmatrix} 0 & 1 \\ 1 & 0 \end{pmatrix}, \quad \sigma_y = \begin{pmatrix} 0 & -i \\ i & 0 \end{pmatrix}, \quad \sigma_z = \begin{pmatrix} 1 & 0 \\ 0 & -1 \end{pmatrix}. \tag{1.23}$$

The spin operators $\widehat{S}_x$, $\widehat{S}_y$ and $\widehat{S}_z$ satisfy the same cyclic commutation relations Eq. (1.17–1.19) as the angular momentum operators $\widehat{L}_x$, $\widehat{L}_y$ and $\widehat{L}_z$. However, there is no way the spin variables can be represented as differential operators acting on wave functions.

Other examples of quantum ideas that cannot be described using wave functions are iso-spin, strangeness, charm *etc.*. The existence of systems of this type requires a significant generalisation of the quantum formalism, so as to be applicable to these more complex situations whilst reproducing the familiar results of elementary wave mechanics. The explication of such a formalism is one of the main goals of this course.

Chapter 2

VECTOR SPACES

2.1 Representations of Probabilities

2.1.1 The Mathematical Representation of Classical Probabilities

Two of the basic features of wave mechanics are that (i) two wave functions can be added together to give a new wave function for the system; and (ii) observables are represented by (differential) operators that are linear, and which therefore respect this superposition principle. However, there are many examples of quantum-mechanical systems whose states *cannot* be represented by wave functions, and even systems whose states can be so represented also admit representations that are not of this type. Nevertheless, in all cases there is always a precise analogue of the central idea of linear superposition, and this is coded mathematically in the concept of a *vector space*.

The fundamental importance of vector spaces for theoretical physics can be seen by thinking in general about what types of mathematical structure might be used to represent probabilities. For example, consider some physical system that has an observable A whose possible values range over a finite set $a_1, a_2, \ldots, a_N$. What options are open to us when trying to construct a theory that associates, to any state s of the system, a set of probabilities $\text{Prob}(A = a_i; s)$, $i = 1, 2, \ldots, N$, for obtaining the result

a_i when measuring[1] A?

For any such theory, the collection of real numbers $\text{Prob}(A = a_i; s)$, $i = 1, 2, \ldots, N$, must satisfy the two crucial conditions:

$$0 \leq \text{Prob}(A = a_i; s) \leq 1 \tag{2.1}$$

and

$$\sum_{i=1}^{N} \text{Prob}(A = a_i; s) = 1. \tag{2.2}$$

The first requirement arises because, in the usual *relative-frequency* interpretation, the probability of getting a particular result is regarded as the limiting proportion of times it is obtained if the experiment is repeated a large number of times (with the system being prepared to be in the same state s before each measurement) and, of course, such a number always lies between 0 and 1. Note that if the value a_i can be guaranteed *never* to be found, then $\text{Prob}(A = a_i; s) = 0$. On the other hand, if a_i is certain to be found, then $\text{Prob}(A = a_i; s) = 1$. Values of $\text{Prob}(A = a_i; s)$ lying between 0 and 1 correspond to the situation in which the result of measuring A might be a_i, but it could also be one of the other allowed values. The second condition Eq. (2.2) reflects mathematically the obvious requirement that the probability of obtaining *some* result must be 1.

The classical way of constructing numbers that satisfy Eq. (2.1–2.2) involves the use of 'generalised volumes' or, more precisely, measure theory. For example, suppose that X is some N-dimensional region in a flat space with a set of Cartesian coordinates $x^1, x^2, \ldots, x^N$. Then the standard volume of X is defined as[2]

$$\text{vol}(X) := \int \ldots \int_X dx^1\, dx^2 \ldots dx^N, \tag{2.3}$$

where the multiple integral is taken over the region X. If X_1 and X_2 are two such regions that are disjoint, then it follows from the definition of

---

[1]Expressing things in terms of the results of measurements is rather instrumentalist in character. A more realist philosophy would associate a state with the probability that the quantity A *has* the value a_i at some specified time. These two forms are almost synonymous in classical physics. However, as we shall see, a sharp distinction must be drawn between them in quantum theory.

[2]The notation $A := B$ is not an equation relating A to B. It means that the object on the left (A) is *defined* by the expression (B) appearing on the right.

integration that this volume assignment is additive:

$$\mathrm{vol}(X_1 \bigcup X_2) = \mathrm{vol}(X_1) + \mathrm{vol}(X_2). \tag{2.4}$$

In particular, this implies that $\mathrm{vol}(X_1 \cup X_2)/\mathrm{vol}(X) = \mathrm{vol}(X_1)/\mathrm{vol}(X) + \mathrm{vol}(X_2)/\mathrm{vol}(X)$ (assuming that the volumes are all finite).

Now suppose that X is divided into disjoint pieces $X_1, X_2, \ldots, X_N$ whose union is X and whose (finite) volumes are denoted $V_1, V_2, \ldots, V_N$ respectively. Then the additivity property of the integral shows that $\sum_{i=1}^N V_i = V$ (the volume of X), and hence $\sum_{i=1}^N V_i/V = 1$. Furthermore, since X_i is a subset of X, $0 \leq V_i \leq V$, and hence $0 \leq V_i/V \leq 1$. Thus, if we set $\mathrm{Prob}(A = a_i; s)$ equal to V_i/V, the two key conditions Eq. (2.1) and Eq. (2.2) are satisfied automatically.

The important ingredient here is the additivity property of the integral: it is this that makes feasible the association of a probability with $\mathrm{vol}(X_i)/\mathrm{vol}(X)$. However, this property is preserved if we define a 'generalised volume'

$$\mathrm{vol}_\rho(X) := \int \cdots \int_X \rho(x^1, x^2, \ldots, x^N) dx^1\, dx^2 \ldots dx^N, \tag{2.5}$$

where ρ is any positive, integrable function on X. Nothing is lost by normalising ρ so that $\mathrm{vol}_\rho(X) = 1$, in which case we can model a probability by the definition

$$\mathrm{Prob}(A = a_i; \rho) = \mathrm{vol}_\rho(X_i) = \int \cdots \int_{X_i} \rho(x^1, x^2, \ldots, x^N) dx^1\, dx^2 \ldots dx^N. \tag{2.6}$$

Note that, in this example, the 'state' of the system is represented mathematically by the *probability density* function ρ. Different choices of the space X and the function ρ give different models for the probability distribution of A. This idea can be extended to the more generalised concept of a 'probability measure' that we shall mention briefly in our discussion in Section 4.3 of the use of probability in classical physics.

2.1.2 The Origin of Quantum Probabilities

The technique discussed above for constructing probabilities is of major importance, and provides the mathematical foundation for *all* normal uses

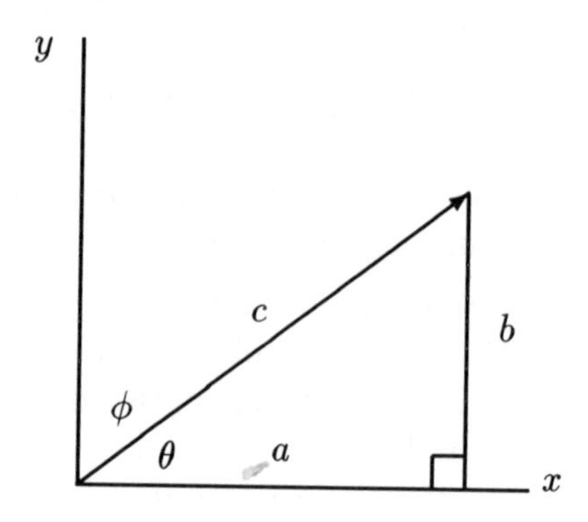

Figure 2.1: The Pythagoras theorem in 2 dimensions

of probability theory, including the whole of classical statistical physics (and hence thermodynamics). However, the crucial feature of the probabilities used in quantum theory is that they do *not* arise in this way but rather from something quite different, namely the Pythagoras theorem.

To illustrate the main idea, consider first the sketch in Figure 2.1 of the right-angled triangle made by a vector in a two-dimensional space. The Pythagoras theorem says that $a^2+b^2=c^2$, which implies $\left(\frac{a}{c}\right)^2+\left(\frac{b}{c}\right)^2=1$. This is just $\cos^2\theta+\sin^2\theta=1$ or, equivalently,

$$\cos^2\theta+\cos^2\phi=1 \tag{2.7}$$

where θ and ϕ are the angles made by the vector with the x and y axes respectively. However, the square of the cosine of any angle necessarily lies between 0 and 1, and so the result Eq. (2.7) shows that a model for a probability distribution for an observable A with only two possible values a_1 and a_2 can be obtained by associating $\mathrm{Prob}(A=a_1;s)$ and $\mathrm{Prob}(A=a_2;s)$ with $\cos^2\theta$ and $\cos^2\phi$ respectively. Different probability distributions (corresponding, for example, to different states s of the system, or to a different observable) can be obtained by changing either the vector or the pair of orthogonal vectors used to define the x and y axes.

The example above can be extended at once to three dimensions. More precisely, let θ_1, θ_2 and θ_3 denote the angles made by a unit vector $\mathbf{v}$ with the three coordinate axes in the familiar three-dimensional space of Newtonian physics; *i.e.*, $\cos\theta_1=\mathbf{i}\cdot\mathbf{v}$, $\cos\theta_2=\mathbf{j}\cdot\mathbf{v}$ and $\cos\theta_3=\mathbf{k}\cdot\mathbf{v}$, where **i**, **j** and **k** are the usual unit vectors. Then a well-known result (used, for example, in the theory of Miller indices in crystal lattices) is that

$$\cos^2\theta_1 + \cos^2\theta_2 + \cos^2\theta_3 = 1. \tag{2.8}$$

Thus we now have a model for the probabilities of an observable that has *three* possible values.

This construction can be generalised to any number of dimensions—including, with some care, an infinite number—and constitutes the essence of the general mathematical framework of quantum theory (albeit using vectors whose components are complex, rather than real, numbers). Thus states are represented by vectors in a vector space, and to each observable there corresponds a set of vectors that are an appropriate generalisation of **i**, **j** and **k**. Each one of these special vectors corresponds to a particular value of the observable, and the probability of obtaining that value is given by the (complex analogue) of the cosine-squared of the angle between this vector and the state vector. It is a salutary thought that the heart of the radically different natures of classical and quantum probability is just the difference between numbers obtained from ratios of volumes, and numbers that come from the Pythagoras theorem!

2.2 From Wave Functions to State Vectors

As was mentioned briefly in Chapter 1, the spin states of an electron are represented by column matrices $\binom{a}{b}$, which we now wish to view as elements of some vector space. This poses the important question of what is meant by a linear combination $\alpha\binom{a_1}{a_2} + \beta\binom{b_1}{b_2}$ of a pair of such matrices $\binom{a_1}{a_2}$ and $\binom{b_1}{b_2}$, where α and β are complex numbers. Operations of this type must exist if we are to preserve the all-important superposition property of quantum states. In effect, we are asking for the appropriate definitions of (i) multiplying a column matrix by a complex number; and (ii) taking the sum of two column matrices.

It must be emphasised that the operative word is 'definition': there is no *a priori* meaning of a linear combination of column matrices. This need for a preliminary definition of mathematical concepts is rather general, and arises already in wave mechanics. A state of the system there is represented by a *function*—*i.e.*, by a *map* from the real numbers (for motion in one dimension) into the complex numbers—and, strictly speaking, it is necessary to define what is meant by a linear combination of such maps. The fact that the linear operations on wave functions need to be *defined* is usually not mentioned in a first course on wave mechanics, but it is an important step along the path to appreciating the mathematical framework of general quantum theory.

The crucial wave-function definitions are as follows. If ψ is such a map and if[3] $\lambda \in \mathbb{C}$, the new map, denoted $\lambda\psi$, is defined if we give its values on every $x \in \mathbb{R}$ (the set of all real numbers), and similarly for the sum of two functions ψ and ϕ. The appropriate definitions (which are used implicitly in introductory wave mechanics) are

$$(\lambda\psi)(x) := \lambda\psi(x) \tag{2.9}$$

and

$$(\psi + \phi)(x) := \psi(x) + \phi(x) \tag{2.10}$$

for all $x \in \mathbb{R}$. It should be noted that the operations used in the right-hand sides of these two definitions involve only the familiar laws of multiplication and addition of complex numbers.

To obtain some idea of what might be meant by a linear combination of column matrices, let us consider a special situation in wave mechanics where we are interested only in those states whose wave functions are linear combinations of some particular pair u_1, u_2 of eigenfunctions of some self-adjoint differential operator $\widehat{M}$ (for example, this might be a system with two energy levels, such as is often invoked to give a simple model of laser action). Thus

$$\begin{aligned} \widehat{M}u_1(x) &= m_1 u_1(x) \\ \widehat{M}u_2(x) &= m_2 u_2(x) \end{aligned} \tag{2.11}$$

where m_1 and m_2 are the corresponding eigenvalues, with $m_1 \neq m_2$. This model could be used with any other operator $\widehat{N}$ that has the special property that both $\widehat{N}u_1$ and $\widehat{N}u_2$ are linear combinations of u_1 and u_2 only.

The most general (non-normalised) wave function involving these states can be written as $\psi(x) = \psi_1 u_1(x) + \psi_2 u_2(x)$, where ψ_1 and ψ_2 are any pair

[3]The notation $x \in X$ means that x is a member of the set X. The symbol $\mathbb{C}$ denotes the set of all complex numbers.

of complex numbers. The values of ψ_1 and ψ_2 completely determine the state ψ, which can therefore be represented unambiguously by the column matrix $\binom{\psi_1}{\psi_2}$. Thus there is a one-to-one correspondence between such column matrices and wave-function states of this type.[4]

For example, if $\psi(x) = \psi_1 u_1(x) + \psi_2 u_2(x)$ and $\phi(x) = \phi_1 u_1(x) + \phi_2 u_2(x)$ then these wave functions are represented by the column matrices $\binom{\psi_1}{\psi_2}$ and $\binom{\phi_1}{\phi_2}$ respectively. However, according to Eq. (2.9), we have $(\lambda\psi)(x) = \lambda(\psi(x)) = \lambda\psi_1 u_1(x) + \lambda\psi_2 u_2(x)$, *i.e.*, $\lambda\psi$ is represented by the pair of complex numbers $\lambda\psi_1$ and $\lambda\psi_2$, and hence by the column matrix $\binom{\lambda\psi_1}{\lambda\psi_2}$. Similarly, Eq. (2.10) shows that $(\psi+\phi)(x) = \psi_1 u_1(x) + \psi_2 u_2(x) + \phi_1 u_1(x) + \phi_2 u_2(x) = (\psi_1 + \phi_1)u_1(x) + (\psi_2 + \phi_2)u_2(x)$, so that $\psi + \phi$ is represented by the pair of numbers $\psi_1 + \phi_1$ and $\psi_2 + \phi_2$, *i.e.*, by the column matrix $\binom{\psi_1+\phi_1}{\psi_2+\phi_2}$. Thus, in this case, the appropriate definitions for the operations on column matrices are

$$\lambda\begin{pmatrix} a_1 \\ a_2 \end{pmatrix} := \begin{pmatrix} \lambda a_1 \\ \lambda a_2 \end{pmatrix} \tag{2.12}$$

and

$$\begin{pmatrix} a_1 \\ a_2 \end{pmatrix} + \begin{pmatrix} b_1 \\ b_2 \end{pmatrix} := \begin{pmatrix} a_1 + b_1 \\ a_2 + b_2 \end{pmatrix}. \tag{2.13}$$

The discussion above can be extended to include any finite set of eigenfunctions $u_1, u_2, \ldots, u_N$, with the wave functions of interest being those that can be expanded in the form $\psi(x) = \sum_{i=1}^{N} \psi_i u_i(x)$. With care, one can even use an infinite set of eigenfunctions once the meaning of the infinite sum $\psi(x) = \sum_{i=1}^{\infty} \psi_i u_i(x)$ has been clarified (see Section 2.5). Since any wave function can always be expanded in terms of the eigenfunctions of a self-adjoint operator, it follows that the states of *any* wave-mechanical system can always be represented by infinite column matrices; this is the form in which Heisenberg originally discovered quantum theory.

In the case of the spin states of an electron, a natural guess for the superposition law is that it also is defined by the equations Eq. (2.12–2.13). However, it should be emphasised once again that, in this case, the combination laws in Eq. (2.12–2.13) cannot be justified by any appeal

[4]To avoid any possible confusion it must be emphasised that there is no physically meaningful way in which the spin states of an electron can be associated like this with a pair of wave functions. Nevertheless, the wave-function example gives a genuine insight into how to perform algebraic operations on column matrices.

to an underlying wave theory but are instead imposed *ab initio* as the appropriate laws for states of this type, rather as—in wave mechanics—the combination laws Eq. (2.9–2.10) are assumed to be the correct choice. Of course, as always in theoretical physics, the validity or otherwise of such assumptions rests ultimately on the empirical effectiveness of the resulting theory.

From a mathematical perspective, our main task is to develop an appropriate framework within which these definitions of the superposition of wave functions or of column matrices can be seen as special cases of some all-embracing structure. This is provided by the ideas of vector space theory.

2.3 The Definition of a Vector Space

2.3.1 The Concept of a Group

The ideas lying behind the general notion of a vector space can be introduced in various ways. One possibility is to extract the essential properties of the specific operations in Eq. (2.9–2.10) and Eq. (2.12–2.13). Another is to base the discussion on the results obtained by manipulating the familiar three-dimensional vectors used in Newtonian mechanics. In either case, it can be seen that the basic idea of a vector space is that

1. there exists a way of combining any pair of vectors $\vec{u}$, $\vec{v}$ to give a new vector written $\vec{u} + \vec{v}$; and

2. to any number λ and vector $\vec{v}$ there is associated a new vector written $\lambda\vec{v}$. (In quantum theory the numbers λ are complex; for Newtonian mechanics they are real.)

However, these combination laws are not arbitrary but are subject to several restrictions which give them their essential *linear* nature.

Combination laws play a central role in many branches of mathematics, one of the most important of which is group theory. In certain types of quantum theory (for example, when considerations of special relativity are taken into account, as in relativistic quantum field theory) a central role is played by groups and their representations. The explicit use of groups has been minimised in these notes, but nevertheless it is worth giving some of the basic ideas.

Definition A *group* is a set G equipped with a 'combination law' that associates with each pair of elements $a, b \in G$ another element, written ab, that satisfies the following three axioms:

1. The combination law is *associative.* That is, for all elements $a, b, c \in G$ we have $a(bc) = (ab)c$.

2. There exists a *unit* element $e \in G$ with the property that, for all $g \in G$, $ge = eg = g$.

3. To each element $g \in G$ there exists an *inverse* element, written g^{-1}, with the property that $gg^{-1} = g^{-1}g = e$.[5]

The group is said to be *abelian* (or *commutative*) if, for all $a, b \in G$, we have $ab = ba$.

Examples

1. The simplest non-trivial group is one with two elements. For example, consider the set whose elements are the pair of matrices $e := \left(\begin{smallmatrix} 1 & 0 \\ 0 & 1 \end{smallmatrix}\right)$ and $a := \left(\begin{smallmatrix} 0 & 1 \\ 1 & 0 \end{smallmatrix}\right)$. This satisfies the axioms for a group if the combination law is defined to be normal matrix multiplication, with the unit element being the matrix $\left(\begin{smallmatrix} 1 & 0 \\ 0 & 1 \end{smallmatrix}\right)$ and with $a^{-1} := a$. The only non-trivial product is

$$\begin{pmatrix} 0 & 1 \\ 1 & 0 \end{pmatrix} \begin{pmatrix} 0 & 1 \\ 1 & 0 \end{pmatrix} = \begin{pmatrix} 1 & 0 \\ 0 & 1 \end{pmatrix}. \tag{2.14}$$

Thus the basic products are $ee = e$, $ea = a$, $ae = a$ and $aa = e$. Clearly this group is abelian.

Now consider the set of the two real numbers[6] $\{1, -1\}$. This is a group with respect to normal multiplication of numbers, with the unit element

[5] It is straightforward to show from these axioms that the unit element is necessarily unique, as is the inverse of any element in the group.

[6] The notation $\{x_1, x_2, \ldots, x_n\}$ means the set whose elements are the objects $x_1, x_2, \ldots, x_n$.

being the number 1 and with the inverse of -1 being -1. The only non-trivial product is $(-1)(-1) = 1$. Hence, if we denote 1 and -1 by e and a respectively, we recover the group law above.

Thus the two sets $\left\{\begin{pmatrix}1&0\\0&1\end{pmatrix}, \begin{pmatrix}0&1\\1&0\end{pmatrix}\right\}$ and $\{1,-1\}$ produce the *same* abstract group structure (with respect to the appropriate combination laws) even though, as sets, they are clearly different. Technically, the two groups are said to be *isomorphic* to each other—a concept that is discussed briefly below. The abstract group of which they are both concrete examples is conventionally denoted $\mathbf{Z}_2$, and is known as the *cyclic group* of order 2.

2. The set $\mathbb{R}_*$ of all non-zero real numbers carries an abelian group law in which the law of combination is multiplication. The unit element is the number 1, and the group inverse of any number $r \in \mathbb{R}_*$ is just its numerical inverse $\frac{1}{r}$.

3. The real numbers $\mathbb{R}$ form an abelian group in which the law of combination is addition. The unit element is the number 0, and the group inverse of any number r is $-r$. Both groups $\mathbb{R}$ and $\mathbb{R}_*$ have obvious analogues using complex numbers.

4. An important example of a *non*-abelian group is provided by the set $GL(N,\mathbb{R})$ (with $N > 1$) of all $N \times N$ real matrices A with the property that $\det A \neq 0$. The combination law is matrix multiplication, *i.e.*, $(AB)_{ij} := \sum_{k=1}^{N} A_{ik}B_{kj}$, $i,j = 1,2,\ldots,N$, and the unit element is the diagonal matrix $\mathbb{1} := \mathrm{diag}(1,1,\ldots,1)$. The group inverse of a matrix is its inverse as a matrix (which is guaranteed to exist by the condition $\det A \neq 0$). This group is known as the *real general linear group*.

Similarly, the *complex linear group* $GL(N,\mathbb{C})$ is defined to be the set of all $N \times N$ complex, invertible matrices.

2.3.2 The Basic Idea of a Vector Space

After the short diversion into group theory we can return to the definition of a vector space.

Definition A (complex)[7] *vector space* is a set V equipped with a

[7] A *real* vector space is defined in the same way but using the real, rather than complex, numbers.

law of combination that associates with each pair of vectors $\vec{u}, \vec{v} \in V$ a third vector, written $\vec{u} + \vec{v}$. There is also a combination law (called *scalar multiplication*) that associates with each $\vec{v} \in V$ and $\lambda \in \mathbb{C}$ a vector, written $\lambda\vec{v}$ (or, less commonly, $\vec{v}\lambda$). These laws are required to satisfy the following axioms:

1. The '+' law makes V into an Abelian group. That is:

   (a) For all $\vec{u}, \vec{v}, \vec{w} \in V$ we have associativity: $\vec{u} + (\vec{v} + \vec{w}) = (\vec{u} + \vec{v}) + \vec{w}$.

   (b) There is a *null* vector $\vec{0}$ (the unit element for the abelian group) that satisfies $\vec{v} + \vec{0} = \vec{0} + \vec{v} = \vec{v}$ for all $\vec{v} \in V$.

   (c) To each vector $\vec{v} \in V$ there exists an inverse element, denoted[8] $-\vec{v}$, such that $\vec{v} + (-\vec{v}) = \vec{0}$.

   (d) For all $\vec{u}, \vec{v} \in V$ we have $\vec{u} + \vec{v} = \vec{v} + \vec{u}$.

2. The '+' law is compatible with the scalar multiplication law, and with the usual additive and multiplicative properties of the complex numbers, in the sense that

$$\alpha(\vec{u} + \vec{v}) = \alpha\vec{u} + \alpha\vec{v} \tag{2.15}$$

$$(\alpha + \beta)\vec{v} = \alpha\vec{v} + \beta\vec{v} \tag{2.16}$$

$$\alpha(\beta\vec{v}) = (\alpha\beta)\vec{v} \tag{2.17}$$

$$1\vec{v} = \vec{v} \tag{2.18}$$

$$0\vec{v} = \vec{0} \tag{2.19}$$

   for all $\vec{u}, \vec{v} \in V$ and $\alpha, \beta \in \mathbb{C}$.

Examples

1. The set of all vectors $\mathbf{v}$ used in Newtonian physics is a real vector space in the sense of the above, with addition of vectors and multiplication by a real number being defined in the familiar way.

---

[8]By convention, the inverse of an element a in an abelian group is usually written $-a$ (rather than a^{-1}) and the 'product' of a, b is written as $a + b$ (rather than ab). It is also conventional to write $a + (-b)$ as $a - b$.

2. The set $\mathbb{C}^N$ of all $N \times 1$ complex column matrices is a complex vector space with respect to the operations

$$\lambda \begin{pmatrix} c_1 \\ \vdots \\ c_N \end{pmatrix} := \begin{pmatrix} \lambda c_1 \\ \vdots \\ \lambda c_N \end{pmatrix} \tag{2.20}$$

and

$$\begin{pmatrix} c_1 \\ \vdots \\ c_N \end{pmatrix} + \begin{pmatrix} d_1 \\ \vdots \\ d_N \end{pmatrix} := \begin{pmatrix} c_1 + d_1 \\ \vdots \\ c_N + d_N \end{pmatrix}, \tag{2.21}$$

with the null vector being

$$\vec{0} = \begin{pmatrix} 0 \\ \vdots \\ 0 \end{pmatrix}. \tag{2.22}$$

This is just the $\mathbb{C}^N$ version of the 2×1 example in Eq. (2.12–2.13).

Similarly, the set $\mathbb{R}^N$ of all $N \times 1$ real column matrices is a real vector space with respect to the analogous operations.

3. The simplest case is $\mathbb{C}^1$, which is just the set of complex numbers. This is a vector space in which the combination law is normal addition. Similarly, scalar multiplication is just ordinary multiplication of complex numbers.

4. The set $\mathbb{C}^\infty$ of all infinite sequences of complex numbers $(c_1, c_2, \ldots)$ is a vector space whose operations are the obvious analogues of those in Eq. (2.20–2.21).

5. The set $M(N, \mathbb{C})$ of all $N \times N$ complex matrices is a vector space with respect to the operations:

$$\lambda \begin{pmatrix} c_{11} & c_{12} & \ldots & c_{1N} \\ c_{21} & c_{22} & \ldots & c_{2N} \\ \vdots & \vdots & \ddots & \vdots \\ c_{N1} & c_{N2} & \ldots & c_{NN} \end{pmatrix} := \begin{pmatrix} \lambda c_{11} & \lambda c_{12} & \ldots & \lambda c_{1N} \\ \lambda c_{21} & \lambda c_{22} & \ldots & \lambda c_{2N} \\ \vdots & \vdots & \ddots & \vdots \\ \lambda c_{N1} & \lambda c_{N2} & \ldots & \lambda c_{NN} \end{pmatrix} \tag{2.23}$$

and

$$\begin{pmatrix} c_{11} & c_{12} & \dots & c_{1N} \\ c_{21} & c_{22} & \dots & c_{2N} \\ \vdots & \vdots & \ddots & \vdots \\ c_{N1} & c_{N2} & \dots & c_{NN} \end{pmatrix} + \begin{pmatrix} d_{11} & d_{12} & \dots & d_{1N} \\ d_{21} & d_{22} & \dots & d_{2N} \\ \vdots & \vdots & \ddots & \vdots \\ d_{N1} & d_{N2} & \dots & d_{NN} \end{pmatrix} :=$$

$$\begin{pmatrix} c_{11}+d_{11} & c_{12}+d_{12} & \dots & c_{1N}+d_{1N} \\ c_{21}+d_{21} & c_{22}+d_{22} & \dots & c_{2N}+d_{2N} \\ \vdots & \vdots & \ddots & \vdots \\ c_{N1}+d_{N1} & c_{N2}+d_{N2} & \dots & c_{NN}+d_{NN} \end{pmatrix}. \quad (2.24)$$

A similar definition applies to the real vector space $M(N, \mathbb{R})$ of real, $N \times N$ matrices. Both definitions can readily be extended to include *rectangular* matrices.

6. If X is any set and V is any vector space, then the set $\mathrm{Map}(X, V)$ of all maps from X into V is a vector space with the operations

$$(\lambda\psi)(x) := \lambda(\psi(x)) \quad (2.25)$$

and

$$(\psi + \phi)(x) := \psi(x) + \phi(x) \quad (2.26)$$

for all $x \in X$, for all $\psi, \phi \in \mathrm{Map}(X, V)$, and all $\lambda \in \mathbb{C}$. Note that the vector space structure of V appears explicitly in the right hand sides of Eq. (2.25) and Eq. (2.26).

The simplest example is when $V = \mathbb{C}$, a special case of which (when $X = \mathbb{R}$) has been considered already in Eq. (2.9–2.10). If we choose $X = \mathbb{R}^N$, an element of $\mathrm{Map}(\mathbb{R}^N, \mathbb{C})$ is a map from $\mathbb{R}^N$ into $\mathbb{C}$, and can therefore be regarded as a complex-valued function of N real variables.[9] In particular, the set of all wave functions $\psi : \mathbb{R}^N \to \mathbb{C}$ in elementary wave mechanics in N dimensions carries the natural structure of a vector space. From this perspective, the use of vector spaces in general quantum theory is a natural extension of the mathematical structure of wave mechanics.

[9]Note that the vector space structure of $\mathbb{R}^N$ plays no role in this construction. The only feature of $\mathbb{R}^N$ that is used is that each element corresponds to a collection $(r_1, r_2, \dots, r_N)$ of N real numbers.

2.3.3 Linear Maps and Isomorphisms

A general, and very important, concept in mathematics is that of a *morphism*. This is a map between two structures of the same type (two groups, two vector spaces *etc.*) that is compatible with their underlying structure. For example, a morphism between a pair of groups G_1 and G_2 is any map[10] $\phi : G_1 \to G_2$ such that $\phi(ab) = \phi(a)\phi(b)$ for all $a, b \in G_1$. (In the context of group theory, such a map is usually called a *homomorphism*.) A special, and important, case is when ϕ is also a *bijection*[11] between the groups. The groups are then said to be *isomorphic* to each other, and constitute different concrete manifestations of the same abstract group law. For example, the map which sends $1 \mapsto \begin{pmatrix} 1 & 0 \\ 0 & 1 \end{pmatrix}$ and $-1 \mapsto \begin{pmatrix} 0 & 1 \\ 1 & 0 \end{pmatrix}$ is an isomorphism between the two examples of a $\mathbf{Z}_2$ group mentioned earlier.

In the context of vector spaces, a morphism is called a linear map. More precisely:

Definition A *linear map* between two vector spaces V_1 and V_2 is a map $L : V_1 \to V_2$ which is compatible with the vector space structure in the sense that

$$L(\alpha\vec{u} + \beta\vec{v}) = \alpha L(\vec{u}) + \beta L(\vec{v}) \tag{2.27}$$

for all $\alpha, \beta \in \mathbb{C}$ and $\vec{u}, \vec{v} \in V_1$.

The map is *anti-linear* if

$$L(\alpha\vec{u} + \beta\vec{v}) = \alpha^* L(\vec{u}) + \beta^* L(\vec{v}) \tag{2.28}$$

for all complex numbers α, β and vectors $\vec{u}, \vec{v}$ in V_1.

If a linear map $L : V_1 \to V_2$ is a bijection, then L is called an *isomorphism* between V_1 and V_2, and V_1 and V_2 are said to be *isomorphic*, denoted by $V_1 \simeq V_2$. This is an important concept since—as in the case of groups—isomorphic spaces can be regarded as being identical to each

[10]The conventional notation for indicating that f is a map from a set X to a set Y is to write $f : X \to Y$. If the element $x \in X$ is taken by f to the element $y \in Y$ then one writes $x \mapsto y$.

[11]In general, a map f from a set X to a set Y is said to be a *bijection* if it is *one-to-one* (*i.e.*, $f(x_1) = f(x_2)$ implies $x_1 = x_2$) and *onto* (*i.e.*, for any $y \in Y$ there exists an $x \in X$ such that $y = f(x)$). Thus a bijection sets up a one-to-one correspondence between the elements of X and the elements of Y. It has a unique *inverse*, denoted f^{-1}, that maps Y onto X and which takes $y \in Y$ to the (unique) point $x \in X$ such that $y = f(x)$.

other *qua* vector spaces, even if the underlying sets are not the same concrete mathematical object.

For example, there is an isomorphism between the vector spaces $\mathbb{C}^4$ and $M(2,\mathbb{C})$ given by the map

$$\begin{pmatrix} c_1 \\ c_2 \\ c_3 \\ c_4 \end{pmatrix} \mapsto \begin{pmatrix} c_1 & c_2 \\ c_3 & c_4 \end{pmatrix}. \tag{2.29}$$

Clearly this can be generalised to an isomorphism between $\mathbb{C}^{N^2}$ and $M(N,\mathbb{C})$.

Comments

1. The following two general results are useful and easy to prove:

- If $V_1 \simeq V_2$ then $V_2 \simeq V_1$.
- If $V_1 \simeq V_2$ and $V_2 \simeq V_3$ then $V_1 \simeq V_3$.

2. The set of all linear maps $L : V_1 \to V_2$ is itself a vector space. The sum of two maps L_1 and L_2 is defined as

$$(L_1 + L_2)(v) := L_1(v) + L_2(v) \tag{2.30}$$

for all $v \in V_1$. If $\lambda \in \mathbb{C}$, and if L is a linear map from V_1 to V_2, then the product λL is defined to be the linear map

$$(\lambda L)(v) := \lambda L(v) \tag{2.31}$$

for all $v \in V_1$.

A special case is when the second space is just the complex numbers $\mathbb{C}$. The vector space of all linear maps from a complex vector space V to $\mathbb{C}$ is called the *dual* of V, and is denoted V^*.

2.3.4 Subspaces of a Vector Space

The final concept to be discussed briefly in this section is that of a linear subspace W of a vector space V. It is clear what this means intuitively, and the formal definition is that a subset $W \subset V$ is a *linear subspace* of V if it is 'closed' under addition and scalar multiplication. That is, if $\vec{w}_1, \vec{w}_2$ are a pair of vectors that lie in the subset W then so does any linear combination $\alpha\vec{w}_1 + \beta\vec{w}_2$.

Examples

1. In $\mathbb{C}^N$, the set of all N-tuples of complex numbers of the form $(a_1, a_2, \ldots, a_M, 0, 0, \ldots, 0)$ with $M < N$ is a linear subspace which is isomorphic to the vector space $\mathbb{C}^M$.

Similarly, for any finite N the vector space $\mathbb{C}^N$ can be viewed as a linear subspace of the vector space $\mathbb{C}^\infty$ of all infinite sequences of complex numbers.

2. A rather important linear subspace of $\mathbb{C}^\infty$ is the set ℓ^2 which is defined to be all those sequences $(a_1, a_2, \ldots)$ of complex numbers that are *square-summable*:

$$\sum_{i=1}^{\infty} |a_i|^2 < \infty. \tag{2.32}$$

3. An important linear subspace of $\mathrm{Map}(\mathbb{R}^n, \mathbb{C})$ is the space $\mathcal{L}^2(\mathbb{R}^n)$ which is defined to be the subset of all functions ψ that are square integrable:

$$\int_{-\infty}^{\infty} \cdots \int_{-\infty}^{\infty} |\psi(x^1, x^2, \ldots, x^n)|^2 \, dx^1 \, dx^2 \, \ldots \, dx^n \; < \infty. \tag{2.33}$$

This is the space of quantum states for a non-relativistic particle moving in n spatial dimensions.

Note that if W_1 and W_2 are a pair of linear subspaces of V, then their set-theoretic intersection $W_1 \bigcap W_2$ is also a linear subspace; in fact, it is the largest linear subspace of V that is contained in both W_1 and W_2.

On the other hand, the union $W_1 \bigcup W_2$ is *not* a linear subspace since it is not closed under addition of vectors. This can be remedied by defining $W_1 + W_2$ to be the set of all linear combinations of vectors in W_1 and W_2 (the so-called *linear span* of $W_1 \bigcup W_2$). This is the smallest linear subspace of V that contains both W_1 and W_2. As we shall see, these operations play an important role in the interpretation of quantum theory in the context of quantum logic.

2.4 Basis Vectors

2.4.1 The Key Definitions

In the elementary vector calculus used in mechanics, it is frequently useful to employ basis vectors **i**, **j**, **k** in terms of which every vector **v** can be expanded as $\mathbf{v} = v_x\mathbf{i} + v_y\mathbf{j} + v_z\mathbf{k}$ with real coefficients $\{v_x, v_y, v_z\}$. The extension of this idea to a general (complex) vector space is of considerable importance, especially in quantum theory where the basic idea is that an expansion of a wave function ψ as $\psi(x) = \sum_{i=1}^{\infty} \psi_i\, u_i(x)$ can be viewed as an expansion of a vector in terms of basis vectors $u_1, u_2, \ldots$ belonging to a vector space of functions. The key definitions are as follows.

Definition

1. A set of vectors $\{\vec{u_1}, \vec{u_2}, \ldots, \vec{u_N}\}$, $N < \infty$, is *linearly dependent* if there is a set of numbers $\{\alpha_1, \alpha_2, \ldots, \alpha_N\}$ (not all zero) such that

$$\sum_{i=1}^{N} \alpha_i \vec{u_i} = \vec{0}. \tag{2.34}$$

   If there is no such set (or, equivalently, if the only way of satisfying Eq. (2.34) is for all the coefficients α_i to vanish) then the set of vectors is *linearly independent*. This is equivalent to saying that no one of them can be written as a linear combination of the rest.

2. An infinite set of vectors is *linearly independent* if every finite subset of vectors is linearly independent.

3. A vector space is *N-dimensional* (where $N < \infty$) if it contains a subset of N linearly independent vectors, but contains no subset of $N + 1$ such vectors.

   A vector space is said to have an *infinite dimension* if it contains N linearly-independent vectors for each positive integer N.

4. A finite set of N linearly independent vectors in an N-dimensional vector space is called a *basis* set for the space.

Examples

1. For every $N < \infty$, the vector space $\mathbb{C}^N$ has dimension N.

2. The vector spaces $\mathbb{C}^\infty$, ℓ^2 and $\mathcal{L}^2(\mathbb{R}^n)$ all have an infinite dimension.

2.4.2 Expansion Theorems

In general, if S is some set of vectors in a vector space V, the *linear span* of S, denoted $[S]$, is defined to be the set of all finite linear combinations of elements of S; clearly $[S]$ is a linear subspace of V.

The central feature of a basis set $S = \{\vec{e_1}, \vec{e_2}, \ldots, \vec{e_N}\}$ is that $[S] = V$, *i.e.*, any $\vec{v} \in V$ can be expanded in terms of the basis vectors. To see this, note that $\{\vec{v}, \vec{e_1}, \vec{e_2}, \ldots, \vec{e_N}\}$ is a set of $N+1$ vectors, and must therefore be linearly dependent. Thus there exist complex numbers $\alpha_1, \alpha_2, \ldots, \alpha_{N+1}$, not all zero, such that

$$\alpha_1 \vec{e_1} + \alpha_2 \vec{e_2} + \cdots + \alpha_N \vec{e_N} + \alpha_{N+1} \vec{v} = \vec{0}. \tag{2.35}$$

Suppose that $\alpha_{N+1} = 0$. Then Eq. (2.35) becomes

$$\alpha_1 \vec{e_1} + \alpha_2 \vec{e_2} + \cdots + \alpha_N \vec{e_N} = \vec{0} \tag{2.36}$$

and at least one of the coefficients $\alpha_1, \alpha_2, \ldots, \alpha_N$ must be non-zero. But this is inconsistent with the fact that $\{\vec{e_1}, \vec{e_2}, \ldots, \vec{e_N}\}$ is a set of linearly independent vectors, and hence $\alpha_{N+1} \neq 0$. Thus Eq. (2.35) can be solved for $\vec{v}$ as $\vec{v} = -\alpha_{N+1}^{-1} \sum_{i=1}^{N} \alpha_i \vec{e_i}$, which is the desired expansion.

In the general expansion

$$\vec{v} = \sum_{i=1}^{N} v_i \vec{e_i} \tag{2.37}$$

the complex numbers $\{v_1, v_2, \ldots, v_N\}$ are known as the *expansion coefficients* (or *components*) of $\vec{v}$ with respect to the basis set. Note that these expansion coefficients are unique. For suppose that

$$\sum_{i=1}^{N} v_i \vec{e_i} = \sum_{i=1}^{N} v_i' \vec{e_i} \tag{2.38}$$

for some other set of complex numbers $\{v_1', v_2', \ldots, v_N'\}$. Then

$$\sum_{i=1}^{N}(v_i - v_i')\vec{e_i} = \vec{0} \tag{2.39}$$

which, since $\{\vec{e_1}, \vec{e_2}, \ldots, \vec{e_N}\}$ is a linearly-independent set of vectors, implies that $v_i = v_i'$ for all $i = 1, 2, \ldots, N$.

Note that there is only one 'abstract' vector space of any particular finite dimension N. That is, any two N-dimensional spaces V_1 and V_2 are necessarily isomorphic to each other. To see this, it suffices to show that any N-dimensional vector space V is isomorphic to $\mathbb{C}^N$. But if $\{\vec{e_1}, \vec{e_2}, \ldots, \vec{e_N}\}$ is a basis set for V, define a map $i : V \to \mathbb{C}^N$ by $i(\vec{v}) := (v_1, v_2, \ldots, v_N)$. It is easy to prove [Exercise] that this map is an isomorphism. Thus $V_1 \simeq \mathbb{C}^N$ and $V_2 \simeq \mathbb{C}^N$, and hence $V_1 \simeq V_2$.

Examples

1. The vectors $\mathbf{i}, \mathbf{j}, \mathbf{k}$ in ordinary vector calculus are a basis set for the real, three-dimensional space. Another basis set is $\mathbf{i}+\mathbf{j}$, $\mathbf{j}-\mathbf{k}$, $\mathbf{i}+\mathbf{j}-16\mathbf{k}$.

2. An obvious basis set for $\mathbb{C}^2$ is $\vec{e_1} := \binom{1}{0}$ and $\vec{e_2} := \binom{0}{1}$. Any vector $\binom{a}{b}$ can be expanded in terms of these vectors as $\binom{a}{b} = a\binom{1}{0} + b\binom{0}{1} = a\vec{e_1} + b\vec{e_2}$.

Another basis set is $\vec{f_1} := \binom{1}{1}$ and $\vec{f_2} := \binom{1}{-1}$ with the explicit expansion of a general vector $\binom{a}{b}$ being

$$\begin{pmatrix} a \\ b \end{pmatrix} = \frac{(a+b)}{2}\begin{pmatrix} 1 \\ 1 \end{pmatrix} + \frac{(a-b)}{2}\begin{pmatrix} 1 \\ -1 \end{pmatrix}. \tag{2.40}$$

3. More generally, a natural basis set for $\mathbb{C}^N$ is

$$\vec{e_1} := \begin{pmatrix} 1 \\ 0 \\ \vdots \\ 0 \end{pmatrix}, \ \vec{e_2} := \begin{pmatrix} 0 \\ 1 \\ \vdots \\ 0 \end{pmatrix}, \ldots, \ \vec{e_N} := \begin{pmatrix} 0 \\ 0 \\ \vdots \\ 1 \end{pmatrix}. \tag{2.41}$$

From introductory wave mechanics we know that a wave function can be expanded in terms of the eigenfunctions of a self-adjoint operator as

$$\psi(x) = \sum_{i=1}^{\infty} \psi_i \, u_i(x) \tag{2.42}$$

and, as remarked earlier, it is tempting to regard Eq. (2.42) as an infinite-dimensional version of the expansion of a vector in terms of a basis set. This is in fact possible, but some care is needed with the precise meaning of the infinite sum in Eq. (2.42). We shall consider this problem briefly at the end of the next section.

2.5 Scalar Products

2.5.1 Motivation for the Idea

A central role in the probabilistic interpretation of standard wave mechanics is played by the overlap function, defined for any $\psi, \phi \in \mathcal{L}^2(\mathbb{R})$ as

$$\langle \psi, \phi \rangle := \int_{-\infty}^{\infty} \psi^*(x)\phi(x)\, dx. \tag{2.43}$$

A key step in developing the mathematical framework of general quantum theory is to find the analogue of this construction when the states of the system belong to a general vector space. This involves a generalisation of the well-known dot product $\mathbf{u} \cdot \mathbf{v}$ between a pair of vectors $\mathbf{u}$, $\mathbf{v}$ in mechanics. Of course, $\mathbf{u} \cdot \mathbf{v}$ is proportional to the cosine of the angle between $\mathbf{u}$ and $\mathbf{v}$, and hence the use of $\mathbf{u} \cdot \mathbf{v}$ in determining probabilities is in accord with the general motivational remarks in Section 2.1.

In the case of the vector space $\mathbb{C}^N$, we can follow a path analogous to that used in Section 2.2 by considering first the special situation in which $\mathbb{C}^N$ occurs as the state space in wave mechanics. This arises when our attention is restricted to wave functions that can be written in the form $\psi(x) = \sum_{i=1}^{N} \psi_i \, u_i(x)$, where $\{u_1, u_2, \ldots, u_N\}$ is a set of eigenfunctions of some self-adjoint operator whose eigenvalues are non-degenerate. Thus the state ψ is identified with the associated column[12] matrix $(\psi_1, \psi_2, \ldots, \psi_N)^T$

[12]For typographical reasons I have written this column matrix as the transpose of a row matrix.

in $\mathbb{C}^N$. Then, if $\psi(x) = \sum_{i=1}^N \psi_i\, u_i(x)$ and $\phi(x) = \sum_{j=1}^N \phi_j\, u_j(x)$ are a pair of such functions, we have

$$\langle \psi, \phi \rangle = \sum_{i,j=1}^{N} \int_{-\infty}^{\infty} \psi_i^* u_i^*(x) \phi_j u_j(x)\, dx. \tag{2.44}$$

But, since for $i \neq j$ the eigenvalues a_i and a_j are different, we know that

$$\int_{-\infty}^{\infty} u_i^*(x) u_j(x)\, dx = \delta_{ij} \tag{2.45}$$

and hence Eq. (2.44) becomes

$$\langle \psi, \phi \rangle = \sum_{i=1}^{N} \psi_i^* \phi_i. \tag{2.46}$$

The discussion above suggests that, in general, if the quantum-mechanical state space of some system is $\mathbb{C}^N$, an appropriate definition for the analogue of the overlap function for states $\vec{a} = (a_1, a_2, \ldots, a_N)^T$ and $\vec{b} = (b_1, b_2, \ldots, b_N)^T$ (a pair of column vectors) might be

$$\langle \vec{a}, \vec{b} \rangle := \sum_{i=1}^{N} a_i^* b_i \tag{2.47}$$

which can be written in matrix form as

$$\langle \vec{a}, \vec{b} \rangle = (a_1^*,\, a_2^*, \ldots, a_N^*) \begin{pmatrix} b_1 \\ b_2 \\ \vdots \\ b_N \end{pmatrix}. \tag{2.48}$$

This is precisely the form that is used when representing the spin states of the electron as elements of $\mathbb{C}^2$.

The definition Eq. (2.47) can be compared with the dot product between a pair of vectors $\mathbf{u}, \mathbf{v}$ in ordinary vector calculus

$$\mathbf{u} \cdot \mathbf{v} = u_x v_x + u_y v_y + u_z v_z. \tag{2.49}$$

It is clear that Eq. (2.49) is a special case of Eq. (2.47) with $n = 3$ and employing the real, rather than complex, numbers. Thus it seems plausible that:

- the familiar overlap function of wave mechanics can usefully be regarded as an analogue on the vector space $\mathcal{L}^2(\mathbb{R})$ of the dot product of elementary vector calculus; and
- some such analogue is the appropriate thing to use on any vector space that serves as the space of quantum states for a physical system.

Both assertions are correct, and the appropriate mathematical concept is that of a scalar product.

2.5.2 The Basic Definitions

Definition A *scalar product* (or *inner product*) on a complex vector space V is an assignment to each pair of vectors $\vec{\psi}, \vec{\phi}$ in V of a complex number $\langle\vec{\psi}, \vec{\phi}\rangle$ satisfying the following conditions:

$$\langle\vec{\psi}, (\alpha_1\vec{\phi}_1 + \alpha_2\vec{\phi}_2)\rangle = \alpha_1\langle\vec{\psi}, \vec{\phi}_1\rangle + \alpha_2\langle\vec{\psi}, \vec{\phi}_2\rangle \quad (2.50)$$

$$\langle\vec{\psi}, \vec{\phi}\rangle^* = \langle\vec{\phi}, \vec{\psi}\rangle \quad (2.51)$$

$$\langle\vec{\psi}, \vec{\psi}\rangle \geq 0 \text{ with } \langle\vec{\psi}, \vec{\psi}\rangle = 0 \text{ only if } \vec{\psi} = \vec{0}. \quad (2.52)$$

Comments

1. Equation Eq. (2.50) is the analogue of the vector calculus relation $\mathbf{u} \cdot (r_1\mathbf{v_1} + r_2\mathbf{v_2}) = r_1\mathbf{u} \cdot \mathbf{v_1} + r_2\mathbf{u} \cdot \mathbf{v_2}$. Similarly, Eq. (2.51) corresponds to the result $\mathbf{u} \cdot \mathbf{v} = \mathbf{v} \cdot \mathbf{u}$, and Eq. (2.52) to the fact that the dot product of a vector $\mathbf{u}$ with itself is always greater than zero unless it is the null vector.

2. Conditions Eq. (2.50–2.51) imply that

$$\langle(\alpha_1\vec{\psi}_1 + \alpha_2\vec{\psi}_2), \vec{\phi}\rangle = \alpha_1^*\langle\vec{\psi}_1, \vec{\phi}\rangle + \alpha_2^*\langle\vec{\psi}_2, \vec{\phi}\rangle. \quad (2.53)$$

It is important to note that a complex number 'comes out of the left' of a scalar product with a complex conjugation, whereas it emerges from the right of the bracket unchanged. A failure to remember this is a common cause of difficulty when attempting to solve problems involving the scalar product.

3. To each vector $\vec{\psi} \in V$ there is associated a map $L_{\vec{\psi}} : V \to \mathbb{C}$ defined by

$$L_{\vec{\psi}}(\vec{\phi}) := \langle \vec{\psi}, \vec{\phi} \rangle. \tag{2.54}$$

The linearity condition Eq. (2.50) in the definition of the scalar product means that $L_{\vec{\psi}}$ is a *linear* map from V to $\mathbb{C}$, and hence belongs to the *dual* vector space V^*. Thus the association $\vec{\psi} \mapsto L_{\vec{\psi}}$ defines a natural map from V into its dual space V^*. Note that, by virtue of Eq. (2.53), $\vec{\psi} \mapsto L_{\vec{\psi}}$ is an *anti-linear* map from V to V^*.

This is an appropriate point at which to introduce the famous "Dirac notation", widely used by theoretical physicists (Dirac 1958). Dirac wrote the scalar product between a pair of vectors as $\langle\psi|\phi\rangle$, and called it a "bracket". He then split the bracket in two, and thought of it as a combination of the "bra" $\langle\psi|$ and the "ket" $|\phi\rangle$, which he then viewed as entities in their own right.

There is no problem with the ket $|\phi\rangle$: it is just an alternative notation for what I have called the vector $\vec{\phi}$. However, a bra is somewhat different. Strictly speaking, it should be regarded as a *map* from the vector space V into the complex numbers, *i.e.,* it is more appropriate to write the Dirac bracket in a function form as $\langle\psi|\,(\,|\phi\rangle)$, which is precisely the complex-valued function of vectors considered above. Thus

$$\langle\psi| \;:= L_{|\psi\rangle}, \tag{2.55}$$

so that a bra really belongs to the *dual* space of V. This raises the obvious question of whether there are any elements of V^* (*i.e.,* linear maps from V to $\mathbb{C}$) that are *not* of the form $L_{|\psi\rangle}$ for some vector $|\psi\rangle$ in V. For a finite-dimensional space, the answer is "no": every linear function is of this type. On the other hand, for most *infinite*-dimensional vector spaces with a scalar product, there exist many elements of the dual that cannot be obtained in this way; *i.e.,* there are 'more' bras than kets. However, in the case of a Hilbert space (the typical space used in quantum theory: see below) it can be shown that *every* complex-valued linear function[13] on V arises in this way. Thus there is a one-to-one correspondence between bras and kets, and $\langle\psi|\phi\rangle$ can be thought of as being either (i) the inner product between vectors $|\psi\rangle$ and $|\phi\rangle$; or (ii) the value of the dual vector

[13]To be more precise, the function has to be *continuous* in a certain sense.

$\langle\psi|$ on the vector $|\phi\rangle$. In what follows, I shall mainly use the former interpretation; in either case, it is important to remember that the Dirac form of Eq. (2.51) is $\langle\psi|\phi\rangle^* = \langle\phi|\psi\rangle$.

Examples

1. Equation Eq. (2.47) defines a genuine scalar product on the vector space $\mathbb{C}^N$. This can be extended to the space ℓ^2 of square-summable sequences (see Eq. (2.32)) by

$$\langle\vec{a},\vec{b}\rangle := \sum_{i=1}^{\infty} a_i^* b_i \tag{2.56}$$

where $\vec{a}$ and $\vec{b}$ are now the infinite sequences $\vec{a} = (a_1, a_2, \ldots)$ and $\vec{b} = (b_1, b_2, \ldots)$. However, in order that Eq. (2.56) be a sensible definition it is necessary that the conditions

$$\sum_{i=1}^{\infty} |a_i|^2 < \infty \quad \text{and} \quad \sum_{i=1}^{\infty} |b_i|^2 < \infty \tag{2.57}$$

should be sufficient to guarantee that the sum in Eq. (2.56) converges absolutely. That this is so follows from the famous Schwarz inequality which says that, for any pair of vectors $\vec{\psi}, \vec{\phi}$ in a vector space with an inner product, $|\langle\vec{\psi},\vec{\phi}\rangle| \leq \langle\vec{\psi},\vec{\psi}\rangle^{\frac{1}{2}}\langle\vec{\phi},\vec{\phi}\rangle^{\frac{1}{2}}$, with equality holding if, and only if, $\vec{\psi}$ and $\vec{\phi}$ are proportional to each other. The proof of this important result is given in Section 7.3.

2. A scalar product can be defined [Exercise] on the vector space $M(N,\mathbb{C})$ (see Eq. (2.23–2.24)) by[14]

$$\langle A, B\rangle := \operatorname{tr}(A^\dagger B) \tag{2.58}$$

for $A, B \in M(N,\mathbb{C})$.

3. Equation Eq. (2.43) 'almost' defines a scalar product on the space $\mathcal{L}^2(\mathbb{R})$, as does the obvious generalisation to the space $\mathcal{L}^2(\mathbb{R}^n)$ (*cf.* equation Eq. (2.33)) defined by

$$\langle\vec{\psi},\vec{\phi}\rangle := \int_{-\infty}^{\infty} \cdots \int_{-\infty}^{\infty} \psi(x^1, x^2, \ldots, x^n)^* \phi(x^1, x^2, \ldots, x^n)\, dx^1\, dx^2 \ldots dx^n. \tag{2.59}$$

[14]Note that tr denotes the trace of a matrix (the sum of the diagonal elements), and $A^\dagger$ denotes the hermitian conjugate of A (*i.e.*, $(A^\dagger)_{ij} = A^*_{ji}$ for all $i, j = 1, 2, \ldots, N$).

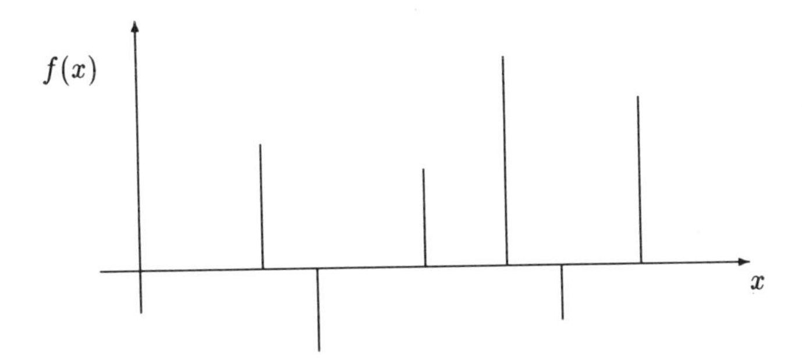

Figure 2.2: A function with zero-width spikes

The qualification 'almost' arises because condition Eq. (2.52) is not strictly true: there exist functions $f(x)$ such that $\int_{-\infty}^{\infty} |f(x)|^2 \, dx = 0$ but for which it is not the case that $f(x) = 0$ for all x. For example, the function in Figure 2.2 is of this type if the spikes have zero width. In a rigorous treatment of the vector spaces $\mathcal{L}^2(\mathbb{R}^n)$ it is necessary to *identify* functions that differ by an 'almost zero' function of this type[15], but this subtlety will be mainly ignored in what follows.

2.5.3 Some Associated Geometrical Concepts

Many of the geometrical concepts in ordinary three-dimensional vector calculus can be generalised to an arbitrary vector space equipped with a scalar product. In particular, we have the following definitions.

Definition

1. The *length* (or *norm*) of a vector is defined to be $\| \vec{\psi} \| := \langle \vec{\psi}, \vec{\psi} \rangle^{\frac{1}{2}}$. Note that, from Eq. (2.52), the norm of a vector is always greater than zero unless the vector itself vanishes.

2. A pair of vectors $\vec{\psi}, \vec{\phi}$ is said to be *orthogonal* if $\langle \vec{\psi}, \vec{\phi} \rangle = 0$.

[15]More precisely, a vector in this space is really an *equivalence class* of functions, in which two functions are defined to be equivalent if they differ by an 'almost zero' function.

3. A set of vectors $\{\vec{u}_1, \vec{u}_2, \ldots, \vec{u}_M\}$ is *orthonormal* if $\langle \vec{u}_i, \vec{u}_j \rangle = \delta_{ij}$ for $i, j = 1, \ldots, M$.

Examples

1. Equation Eq. (2.45) can be interpreted as saying that the set of eigenfunctions of a (non-degenerate) self-adjoint operator is an orthonormal set in the state space $\mathcal{L}^2(\mathbb{R})$ of elementary wave mechanics. Modulo the subtleties involved in handling the infinite-dimensional space, this is in fact a basis set. Note that the 'length' of a function is $\| \vec{\psi} \| = (\int_{-\infty}^{\infty} |\psi(x)|^2 \, dx)^{\frac{1}{2}}$.

2. The column vectors in Eq. (2.41) provide an orthonormal basis set for the vector space $\mathbb{C}^N$. The length of a vector $\vec{a} = (a_1, a_2, \ldots, a_N)$ is $\| \vec{a} \| = \left(\sum_{i=1}^{N} |a_i|^2 \right)^{\frac{1}{2}}$.

Comments

1. Basis sets that are orthonormal are particularly useful since it is possible to compute explicitly the expansion coefficients of any vector. Thus, if $\{\vec{e_1}, \vec{e_2}, \ldots, \vec{e_N}\}$ is an orthonormal basis set for some N-dimensional vector space V, any vector $\vec{\psi}$ in V can be expanded as

$$\vec{\psi} = \sum_{i=1}^{N} \psi_i \, \vec{e_i} \tag{2.60}$$

where the complex numbers $\psi_1, \psi_2, \ldots, \psi_N$ are the expansion coefficients. Now take the scalar product of both sides of Eq. (2.60) with a particular basis vector $\vec{e_j}$. Then

$$\langle \vec{e_j}, \vec{\psi} \rangle = \langle \vec{e_j}, \sum_{i=1}^{N} \psi_i \vec{e_i} \rangle = \sum_{i=1}^{N} \psi_i \langle \vec{e_j}, \vec{e_i} \rangle = \sum_{i=1}^{N} \psi_i \, \delta_{ji} = \psi_j. \tag{2.61}$$

Thus the expansion coefficients are just

$$\psi_j = \langle \vec{e_j}, \vec{\psi} \rangle, \tag{2.62}$$

and we see that the expansion of the vector $\vec{\psi}$ in terms of the orthonormal basis set can be written as

$$\vec{\psi} = \sum_{i=1}^{N} \vec{e_i} \, \langle \vec{e_i}, \vec{\psi} \rangle \tag{2.63}$$

or, in the well-known Dirac form,

$$|\psi\rangle = \sum_{i=1}^{N} |e_i\rangle \langle e_i|\psi\rangle. \tag{2.63'}$$

2. Equation Eq. (2.63) gives a useful result for the scalar product of a pair of vectors $\vec{\phi}, \vec{\psi}$ in terms of their components:

$$\langle \vec{\phi}, \vec{\psi} \rangle = \sum_{i=1}^{N} \phi_i^* \psi_i \tag{2.64}$$

or, in Dirac form,

$$\langle \phi|\psi\rangle = \sum_{i=1}^{N} \langle e_i|\phi\rangle^* \langle e_i|\psi\rangle = \sum_{i=1}^{N} \langle \phi|e_i\rangle\langle e_i|\psi\rangle. \tag{2.64'}$$

In particular,

$$\|\vec{\psi}\|^2 = \langle \vec{\psi}, \vec{\psi} \rangle = \sum_{i=1}^{N} |\psi_i|^2. \tag{2.65}$$

3. The result in Eq. (2.62) is the general analogue of the statement in elementary vector calculus that if $\mathbf{v} = v_x\mathbf{i} + v_y\mathbf{j} + v_z\mathbf{k}$ then $v_x = \mathbf{v} \cdot \mathbf{i}$ *etc.* Similar remarks apply to Eq. (2.64) and Eq. (2.65).

2.5.4 The Expansion Theorems in Infinite Dimensions

Finally, a few comments on how the expansion results above can be generalised to the case where the vector space has an infinite dimension. The main technical problem is to give a proper meaning[16] to the infinite sums that appear in the right hand sides of these expressions. We recall that, in the case of complex numbers, an expression like

$$S = \sum_{i=1}^{\infty} a_i \tag{2.66}$$

[16] An excellent source for rigorous mathematical treatments of many of the topics discussed in these notes is Reed & Simon (1972).

means that the number S is the *limit* of the partial sums $S^M := \sum_{i=1}^{M} a_i$; that is

$$S := \lim_{M \to \infty} \sum_{i=1}^{M} a_i. \tag{2.67}$$

We recall also that a sequence of complex numbers $\alpha_1, \alpha_2, \ldots$ is said to *converge* to a complex number α if, for any positive real number ϵ (however small), there exists an integer n_0 (generally dependent on ϵ) such that $n > n_0$ implies $|\alpha - \alpha_n| < \epsilon$. Put more geometrically, any disk centered on α eventually 'traps' the sequence of numbers.

In all branches of mathematics, infinite sums are invariably defined in this way and hence, in the case of interest, we must decide first what it means to say that a sequence of vectors $\vec{v}_1, \vec{v}_2, \ldots$ in V converges to a limit vector $\vec{v}$.

For a finite-dimensional vector space it seems intuitively clear that what is wanted is that each of the *components* of the vectors $\vec{v}_m$ should converge (in the sense of complex numbers) to the components of $\vec{v}$. However, such a definition would be circular in the infinite-dimensional case since a definition of convergence is needed in the first place to make sense of expanding a vector in terms of a basis set.

It transpires that the most useful concept is to define convergence using a norm on V as an analogue of what is done for a sequence of complex numbers. More precisely, we have:

Definition A sequence of vectors $\vec{v}_m \in V$ is said to *converge strongly* to $\vec{v} \in V$ (written $\vec{v}_m \to \vec{v}$) if the norms $\| \vec{v} - \vec{v}_m \|$ converge to 0 as a sequence of real numbers, *i.e.*, $\lim_{m\to\infty} \| \vec{v} - \vec{v}_m \| = 0$. Thus, for any $\epsilon > 0$, there exists an integer n_0 (which can depend on ϵ) such that $n > n_0$ implies $\| \vec{v} - \vec{v}_n \| < \epsilon$. More geometrically, any ball centered on $\vec{v}$ eventually 'traps' the sequence of vectors.

Comments

1. This definition includes the one given above for the special case where the vector space V is just $\mathbb{C}$ with $\| \alpha \| = |\alpha|$.

2. It can be shown [Exercise] that $\vec{v}_m \to \vec{v}$ implies that the sequence of real numbers $\| \vec{v}_m \|$ converges to $\| \vec{v} \|$, *i.e.*, $\| \vec{v} \| = \lim_{m\to\infty} \| \vec{v}_m \|$.

3. An important consequence of the definition of a scalar product is that it is *continuous* in the sense that, if $\vec{v_m} \to \vec{v}$, then $\langle\vec{\phi},\vec{v}_m\rangle$ converges to $\langle\vec{\phi},\vec{v}\rangle$ for all $\phi \in V$, *i.e.*, $\langle\vec{\phi},\vec{v}\rangle = \lim_{m\to\infty}\langle\vec{\phi},\vec{v}_m\rangle$.

A peculiar problem that can arise in the theory of infinite-dimensional spaces is an analogue of the phenomenon that appears in the case of the real numbers if we restrict ourselves to the subset of rationals. One can have sequences of rational numbers that are clearly 'trying' to converge (*i.e.*, Cauchy convergent[17] sequences), but which do not converge to a rational number *per se*. For example, the decimal expansion of any irrational number (like π) is of this type. In this case, the answer is easy: 'complete' the rationals by adding in the irrational numbers.

Needless to say, the situation for infinite-dimensional vector spaces is more complicated, but many spaces do admit sequences of vectors that are trying hard to converge to something that may not exist as an element of the space. More precisely, we have the following definition.

Definition A sequence of vectors $\vec{v}_1, \vec{v}_2, \ldots$ is said to be a *Cauchy sequence* if the differences between vectors get arbitrarily small, *i.e.*, if for any $\epsilon > 0$, there exists an integer n_0 such that $m, n > n_0$ implies $\| \vec{v}_n - \vec{v}_m \| < \epsilon$.

If any sequence of vectors $\vec{v}_1, \vec{v}_2, \ldots$ converges strongly to $\vec{v}$ then it is easy to show that it is a Cauchy sequence [Exercise]. The crucial question is the converse, *i.e.*, in what types of vector space does every Cauchy sequence converge strongly to a vector in the space?

Definition A scalar product space in which every Cauchy sequence of vectors converges strongly to an element in the space is said to be *complete*. It is then called a *Hilbert space*.

Comments

1. A finite-dimensional vector space is always complete.

2. If V is an incomplete vector space, there is a way of constructing a bigger space $\overline{V}$ that contains V as a linear subspace, and which *is* com-

[17] A sequence of complex (or real) numbers $\alpha_1, \alpha_2, \ldots$ is a *Cauchy convergent sequence* if, for any $\epsilon > 0$, there exists an integer n_0 such that $m, n > n_0$ implies $|\alpha_n - \alpha_m| < \epsilon$.

plete. The process is an exact analogue of the standard way of constructing the real numbers from the rationals.

3. Completeness is a very desirable property since the expansion theorems discussed earlier using finite sums then have exact analogues in the infinite-dimensional case. The only proviso is that the vector space must be *separable*, which means that it has a basis set of vectors that is *countably infinite*—*i.e.*, in the sense of the number of integers—rather than infinite in a 'larger' sense (for example, in the sense of the number of real numbers).

One of the basic structural rules of quantum theory is that the states of a system are represented by vectors in a Hilbert space. The spaces used are almost always separable, and so, from now on, all infinite-dimensional spaces will be assumed to have this property.

In a separable Hilbert space with an orthonormal basis set $\{\vec{e_1}, \vec{e_2}, \dots\}$, the basic expansion expression Eq. (2.60) becomes

$$\vec{\psi} = \sum_{i=1}^{\infty} \psi_i \, \vec{e_i} \tag{2.68}$$

in which the right hand side denotes the strong limit of the partial sums.

4. The vector space ℓ^2 of square-summable sequences is a Hilbert space. This is the space that featured in Heisenberg's original discussion of quantum theory.

5. The space of wave functions $\mathcal{L}^2(\mathbb{R}^n)$ is not complete as it stands. However, it becomes so after a subtle improvement in the concept of integration in which the elementary definition of an integral as the limit of partial Riemann sums has to be replaced with a more sophisticated approach, known as *Lebesgue integration*. It is also necessary to regard two functions as being equivalent if they are equal everywhere except on a subset of the real numbers that has 'Lebesgue measure zero'. All this is implicit in the elementary use of wave functions, although it is rarely spelt out in introductory treatments. We shall denote the final, complete Hilbert space by $L^2(\mathbb{R}^n)$.

6. In the case of wave functions, the statement $\vec{\psi}_m \to \vec{\psi}$ of strong convergence in $L^2(\mathbb{R})$ means that

$$\lim_{m\to\infty} \int_{-\infty}^{\infty} |\psi(x) - \psi_m(x)|^2 \, dx = 0. \tag{2.69}$$

In particular, the expansion Eq. (2.42) means that

$$\lim_{M\to\infty} \int_{-\infty}^{\infty} |\psi(x) - \sum_{i=1}^{M} \psi_i u_i(x)\,|^2\, dx = 0, \tag{2.70}$$

which does *not* imply that the sequence of complex numbers $\sum_{i=1}^{M} \psi_i u_i(x)$ converges to $\psi(x)$ for each $x \in \mathbb{R}$! Convergence of the type in Eq. (2.70) is called *convergence in the mean*, and plays an important role in the classical theory of Fourier analysis.

Chapter 3

LINEAR OPERATORS

3.1 The Basic Definitions

In elementary wave mechanics, states are represented by wave functions, observables are represented by self-adjoint differential operators, and the possible results of a measurement of an observable are the eigenvalues of the corresponding operator. To proceed further we need the analogues of these concepts for a general Hilbert space $\mathcal{H}$.

The notion of a linear operator is straightforward: it is simply a special case of a linear map Eq. (2.27) in which both the domain space (V_1) and the target space (V_2) are the same space $\mathcal{H}$. The concept of *hermiticity* involves the inner product on $\mathcal{H}$. The main definitions are as follows.

Definition

1. A *linear operator* (or just *operator*) $\widehat{A}$ on a Hilbert space $\mathcal{H}$ associates a vector, denoted[1] $\widehat{A}\vec{\psi}$, with every vector $\vec{\psi}$ in $\mathcal{H}$ such that

$$\widehat{A}(\alpha\vec{\psi} + \beta\vec{\phi}) = \alpha\widehat{A}\vec{\psi} + \beta\widehat{A}\vec{\phi} \tag{3.1}$$

for all $\alpha, \beta \in \mathbb{C}$ and $\vec{\psi}, \vec{\phi} \in \mathcal{H}$.

[1] A linear operator is simply a linear map from $\mathcal{H}$ to $\mathcal{H}$. However, it is conventional to denote the effect of acting with $\widehat{A}$ on $\vec{\psi}$ by $\widehat{A}\vec{\psi}$ rather than use the function notation $\widehat{A}(\vec{\psi})$.

2. The *sum* of a pair of operators $\widehat{A}, \widehat{B}$ is the operator $\widehat{A} + \widehat{B}$ defined by
$$(\widehat{A} + \widehat{B})\vec{\psi} := \widehat{A}\vec{\psi} + \widehat{B}\vec{\psi} \tag{3.2}$$
for all $\vec{\psi} \in \mathcal{H}$.

   The *product* of $\widehat{A}$ and $\widehat{B}$ is the operator $\widehat{A}\widehat{B}$ defined by
$$(\widehat{A}\widehat{B})\vec{\psi} := \widehat{A}(\widehat{B}\vec{\psi}) \tag{3.3}$$
for all $\vec{\psi} \in \mathcal{H}$.

   The product of an operator $\widehat{A}$ with a complex number λ is the operator $\lambda\widehat{A}$ defined by
$$(\lambda\widehat{A})\vec{\psi} := \lambda(\widehat{A}\vec{\psi}) \tag{3.4}$$
for all $\vec{\psi} \in \mathcal{H}$.

3. A (non-zero) vector $\vec{u} \in \mathcal{H}$ is an *eigenvector* of $\widehat{A}$ with *eigenvalue a* if
$$\widehat{A}\vec{u} = a\vec{u}. \tag{3.5}$$

4. The set of *matrix elements* of an operator $\widehat{A}$ on a Hilbert space $\mathcal{H}$ is the collection of all numbers $\langle\vec{\psi}, \widehat{A}\vec{\phi}\rangle$ where $\vec{\psi}, \vec{\phi} \in \mathcal{H}$.

5. The *adjoint* (or *hermitian conjugate*) of an operator $\widehat{A}$ is the operator $\widehat{A}^\dagger$ defined by the condition on its matrix elements:
$$\langle\vec{\psi}, \widehat{A}^\dagger\vec{\phi}\rangle = \langle\widehat{A}\vec{\psi}, \vec{\phi}\rangle \tag{3.6}$$
for all $\vec{\psi}, \vec{\phi} \in \mathcal{H}$.

6. An operator $\widehat{A}$ is *self-adjoint* (or *hermitian*) if $\widehat{A} = \widehat{A}^\dagger$. That is, for all $\vec{\psi}, \vec{\phi} \in \mathcal{H}$, the matrix elements of $\widehat{A}$ satisfy the conditions
$$\begin{aligned} \langle\vec{\psi}, \widehat{A}\vec{\phi}\rangle &= \langle\widehat{A}\vec{\psi}, \vec{\phi}\rangle && (3.7) \\ &= \langle\vec{\phi}, \widehat{A}\vec{\psi}\rangle^*. && (3.8) \end{aligned}$$

   Note that, for wave functions, Eq. (3.7) is just the familiar condition Eq. (1.2)
$$\int_{-\infty}^{\infty} \psi^*(x)\,(\widehat{A}\phi)(x)\,dx = \int_{-\infty}^{\infty} (\widehat{A}\psi)^*(x)\,\phi(x)\,dx \tag{3.9}$$
for all square-integrable functions ψ and ϕ.

7. A self-adjoint operator $\widehat{A}$ is *bounded* if its eigenvalues are contained in a finite subspace of the real line.

Comments

1. The set of all linear operators on $\mathcal{H}$ is itself a vector space under the operations defined in Eq. (3.2) and Eq. (3.4).

2. It is possible for an operator to have more than one linearly-independent eigenvector with the *same* eigenvalue. If this is not the case, the operator is said to be *multiplicity-free* or *non-degenerate*.

An eigenvalue a_m of an operator $\widehat{A}$ is said to be *$d(m)$-fold degenerate* if there exist $d(m)$ linearly-independent eigenvectors $\vec{u}_{m1}, \vec{u}_{m2}, \ldots, \vec{u}_{m\,d(m)}$ with this same eigenvalue. Note that, since $\widehat{A}\vec{u}_{mi} = a_m\vec{u}_{mi}$ for all $i = 1, 2, \ldots, d(m)$, it follows that if $c_1, c_2, \ldots, c_{d(m)}$ is any set of $d(m)$ complex numbers, then

$$\widehat{A}\left(\sum_{i=1}^{d(m)} c_i\vec{u}_{mi}\right) = a_m\left(\sum_{i=1}^{d(m)} c_i\vec{u}_{mi}\right), \tag{3.10}$$

so that any linear combination of the eigenvectors $\vec{u}_{m1}, \vec{u}_{m2}, \ldots, \vec{u}_{m\,d(m)}$ is also an eigenvector with the same eigenvalue. Hence the eigenvectors with a given eigenvalue a_m form a *linear subspace* of the vector space with dimension $d(m)$; this is known as the *eigenspace* associated with the eigenvalue a_m. In an infinite-dimensional Hilbert space, it is possible for an eigenvalue to be infinitely degenerate, *i.e.*, $d(m) = \infty$. We shall see an example of this below.

An example of a multiplicity-free operator is the Hamiltonian of the quantised simple harmonic oscillator, each of whose eigenvalues $\frac{1}{2}\hbar\omega, \frac{3}{2}\hbar\omega, \ldots$ (where ω is the angular frequency) is non-degenerate.

An important example in wave mechanics of an operator with degenerate eigenvalues is the total angular momentum $\widehat{L}\cdot\widehat{L} := \widehat{L}_x^2 + \widehat{L}_y^2 + \widehat{L}_z^2$. As mentioned in the Introduction, the possible eigenvalues of $\widehat{L}\cdot\widehat{L}$ are of the form $\ell(\ell+1)\hbar$ with $\ell = 0, 1, 2, \ldots$. The eigenvalue $\ell(\ell+1)$ has a degeneracy $2\ell+1$, and a natural basis set for the associated $(2\ell+1)$-dimensional eigenspace is afforded by the eigenfunctions of the operator $\widehat{L}_z$, whose eigenvalues $m\hbar$ range from $-\ell\hbar$ to $\ell\hbar$ in steps of $\hbar$.

3. In Dirac notation, the matrix element $\langle\vec{\psi}, \widehat{A}\vec{\phi}\rangle$ is written as $\langle\psi|\,\widehat{A}\,|\phi\rangle$. The natural interpretation of this expression is that it is the result of acting

(as a linear function) with the bra $\langle\psi|$ on the ket $\widehat{A}\,|\phi\rangle$. However, it is sometimes useful to think instead of $\widehat{A}$ as 'acting to the left' on $\langle\psi|$ to give a new bra $\langle\psi|\,\widehat{A}$, which then acts on the ket $|\phi\rangle$. In terms of the association of $\langle\chi|$ with $L_{|\chi\rangle}$ in Eq. (2.55), the bra $\langle\psi|\,\widehat{A}$ is the linear map $L_{\widehat{A}^\dagger\,|\psi\rangle}$ from $\mathcal{H}$ to $\mathbb{C}$.

4. When using Dirac notation, it is standard to label the eigenstates by the eigenvalue of the operator. Thus Eq. (3.5) becomes

$$\widehat{A}\,|a\rangle = a\,|a\rangle \tag{3.5$'$}$$

where $|a\rangle$ is called an *eigenket*. But note that this labelling is ambiguous if the eigenvalue is degenerate. In this case, an additional label must be added to the ket to distinguish the different eigenvectors with the same eigenvalue.

The other equations above can also be written in Dirac form. However Eq. (3.7) is rather awkward, and so the definition of self-adjointness is usually written using the (equivalent) version Eq. (3.8) as

$$\langle\psi|\,\widehat{A}\,|\phi\rangle = \langle\phi|\,\widehat{A}\,|\psi\rangle^*. \tag{3.8$'$}$$

Similarly, the defining equation for the adjoint of an operator is

$$\langle\psi|\,\widehat{A}^\dagger\,|\phi\rangle = \langle\phi|\,\widehat{A}\,|\psi\rangle^*. \tag{3.11}$$

Note that if $\widehat{A}\,|a\rangle = a\,|a\rangle$ then, for all $|\psi\rangle$,

$$\langle a|\,\widehat{A}^\dagger\,|\psi\rangle = \langle\psi|\,\widehat{A}\,|a\rangle^* = \langle\psi|a\rangle^* a^* = \langle a|\psi\rangle a^*, \tag{3.12}$$

which means it is consistent to think of the operator $\widehat{A}^\dagger$ as acting on the bra $\langle a|$ by

$$\langle a|\,\widehat{A}^\dagger = a^*\langle a|\,. \tag{3.13}$$

This equation is sometimes useful when manipulating Dirac notation. It can also be derived as

$$\langle a|\,\widehat{A}^\dagger = L_{\widehat{A}\,|a\rangle} = L_{a\,|a\rangle} = a^* L_{|a\rangle} = a^*\langle a| \tag{3.14}$$

where the result $L_{a\,|a\rangle} = a^* L_{|a\rangle}$ follows from the fact that $|\psi\rangle \mapsto L_{|\psi\rangle}$ is an *anti*-linear map from $\mathcal{H}$ to its dual.

5. It must be checked that Eq. (3.6) really does define a linear operator. This presupposes the fact that any operator $\widehat{B}$ is determined uniquely by its matrix elements $\langle\vec{\psi}, \widehat{B}\vec{\phi}\rangle$ for all vectors $\vec{\psi}, \vec{\phi} \in \mathcal{H}$. This in turn rests on the fact that a vector $\vec{\psi}$ is determined uniquely by its scalar products with all other vectors. Delicate problems arise in the case of an infinite-dimensional Hilbert space (for example, there is a subtle difference (Reed & Simon 1972) between an operator being hermitian and being self-adjoint), but we shall judiciously ignore these issues in what follows.

6. Any operator on a finite-dimensional vector space is necessarily bounded. However, there exist plenty of examples of unbounded operators on spaces whose dimension is infinite. For example, the operators $\widehat{x}$ and $\widehat{p}$ in wave mechanics have this property, as do the Hamiltonians of the hydrogen atom and the simple harmonic oscillator.

7. The differential operators of introductory wave mechanics are 'almost' linear operators on the Hilbert space $L^2(\mathbb{R})$. The qualification 'almost' arises because an operator like $-i\frac{d}{dx}$ cannot be defined on *all* vectors in $L^2(\mathbb{R})$ but only on those functions that are differentiable[2]. This is related to the subtlety mentioned above in 5. and is one of the many reasons why, mathematically speaking, 'elementary' wave mechanics is far from being elementary!

8. Let $L : V_1 \to V_2$ be a linear map between a pair of finite-dimensional vector spaces V_1 and V_2 of dimensions N_1 and N_2 respectively. Let $\{\vec{e}_1, \vec{e}_2, \ldots, \vec{e}_{N_1}\}$ and $\{\vec{f}_1, \vec{f}_2, \ldots, \vec{f}_{N_2}\}$ be a pair of basis sets for the two spaces (not necessarily orthonormal), and let $\vec{\psi} = \sum_{i=1}^{N_1} \psi_i \vec{e}_i$ be any vector in V_1. Then, since L is linear, we have

$$L\left(\sum_{i=1}^{N_1} \psi_i\, \vec{e}_i\right) = \sum_{i=1}^{N_1} \psi_i L(\vec{e}_i). \tag{3.15}$$

But each vector $L(\vec{e}_i) \in V_2$ must be expressible as a linear combination of the basis vectors $\{\vec{f}_1, \vec{f}_2, \ldots, \vec{f}_{N_2}\}$ of V_2. Thus there exists a set of complex numbers L_{ji}, $j = 1, 2, \ldots, N_2$, $i = 1, 2, \ldots, N_1$ such that, for each

[2]More precisely, the operator $-i\frac{d}{dx}$ is only defined on those Lebesgue equivalence classes of functions that include a function ψ that is differentiable. The symbol $-i\frac{d\psi}{dx}$ then means the equivalence class of the function obtained by differentiating ψ and multiplying by $-i$.

$i = 1, 2, \ldots, N_1$,

$$L(\vec{e}_i) = \sum_{j=1}^{N_2} \vec{f}_j L_{ji}. \tag{3.16}$$

Substituting Eq. (3.16) into Eq. (3.15) gives

$$\begin{aligned} L(\vec{\psi}) &= \sum_{i=1}^{N_1} \psi_i \left(\sum_{j=1}^{N_2} \vec{f}_j L_{ji} \right) \\ &= \sum_{j=1}^{N_2} \left(\sum_{i=1}^{N_1} L_{ji} \psi_i \right) \vec{f}_j. \end{aligned} \tag{3.17}$$

Thus, if V_1 and V_2 are identified with $\mathbb{C}^{N_1}$ and $\mathbb{C}^{N_2}$ respectively (via the isomorphism that identifies a vector with its set of components with respect to a given basis set), we see that every linear transformation from V_1 to V_2 can be written as the operation of a rectangular matrix in the form

$$\begin{pmatrix} \psi_1 \\ \psi_2 \\ \vdots \\ \psi_{N_1} \end{pmatrix} \mapsto \begin{pmatrix} L_{11} & L_{12} & \ldots & L_{1N_1} \\ L_{21} & L_{22} & \ldots & L_{2N_1} \\ \vdots & \vdots & \ddots & \vdots \\ L_{(N_2-1)1} & L_{(N_2-1)2} & \ldots & L_{(N_2-1)N_1} \\ L_{N_2 1} & L_{N_2 2} & \ldots & L_{N_2 N_1} \end{pmatrix} \begin{pmatrix} \psi_1 \\ \psi_2 \\ \vdots \\ \psi_{N_1} \end{pmatrix} \tag{3.18}$$

which transforms an $N_1 \times 1$-column matrix in $\mathbb{C}^{N_1}$ into an $N_2 \times 1$-column matrix in $\mathbb{C}^{N_2}$.

9. In particular, a linear operator $\widehat{A}$ acting on a single vector space V of dimension N is equivalent to the matrix operation

$$\begin{pmatrix} \psi_1 \\ \psi_2 \\ \vdots \\ \psi_N \end{pmatrix} \mapsto \begin{pmatrix} A_{11} & A_{12} & \ldots & A_{1N} \\ A_{21} & A_{22} & \ldots & A_{2N} \\ \vdots & \vdots & \ddots & \vdots \\ A_{N1} & A_{N2} & \ldots & A_{NN} \end{pmatrix} \begin{pmatrix} \psi_1 \\ \psi_2 \\ \vdots \\ \psi_N \end{pmatrix} \tag{3.19}$$

and, conversely, every such matrix gives rise to a linear operator on V. Thus linear operators on an N-dimensional vector space are in one-to-one correspondence with $N \times N$ square matrices. A particular example is afforded by the famous Pauli spin matrices given in Eq. (1.23), which should be regarded as the matrix representatives of operators acting on the vector space $\mathbb{C}^2$.

Note, however, that the *actual* correspondence between operator and matrix depends on the choice of the basis set. Thus a square matrix can be associated with a specific linear operator only *after* a basis set has been specified. Of course, if the vector space is taken from the beginning to be $\mathbb{C}^N$ (rather than a space isomorphic to it) then Eq. (3.19) can apparently be used unambiguously. In fact, this corresponds to using the natural basis set for $\mathbb{C}^N$ given by Eq. (2.41).

With due care, this idea can be extended to the case of an infinite-dimensional Hilbert space with $n = \infty$. Thus operators can be represented by $\infty \times \infty$ square matrices. This is the formalism within which Heisenberg first discovered quantum theory; it is commonly known as *matrix mechanics*.

10. The matrix representative of an operator $\widehat{A}$ with respect to an *orthonormal* basis set has the explicit form [Exercise!]

$$A_{ij} = \langle \vec{e_i}, \widehat{A}\vec{e_j} \rangle \tag{3.20}$$

or, in Dirac notation,

$$A_{ij} = \langle e_i | \, \widehat{A} \, | e_j \rangle. \tag{3.20$'$}$$

Similarly, the quantities $\langle \vec{\psi}, \widehat{A}\vec{\phi} \rangle$ have the simple matrix expression [Exercise!]

$$\begin{aligned} \langle \vec{\psi}, \widehat{A}\vec{\phi} \rangle &= \sum_{i,j=1}^{N} \psi_i^* A_{ij} \phi_j \\ &= (\psi_1^*, \psi_2^*, \ldots, \psi_N^*) \begin{pmatrix} A_{11} & A_{12} & \ldots & A_{1N} \\ A_{21} & A_{22} & \ldots & A_{2N} \\ \vdots & \vdots & \ddots & \vdots \\ A_{N1} & A_{N2} & \ldots & A_{NN} \end{pmatrix} \begin{pmatrix} \phi_1 \\ \phi_2 \\ \vdots \\ \phi_N \end{pmatrix}. \end{aligned} \tag{3.21}$$

11. On the vector space $\mathbb{C}^N$, the eigenvector equation $\widehat{A}\vec{u} = \lambda\vec{u}$ becomes the matrix equation $(\widehat{A} - \lambda \mathbb{1})\vec{u} = 0$, which admits a solution if and only if $\det(\widehat{A} - \lambda \mathbb{1}) = 0$. Thus the eigenvalues of $\widehat{A}$ are the N roots of this N'th-order polynomial equation in λ (this is called the *characteristic equation* of the matrix).

12. Each pair of vectors $\vec{u}, \vec{v}$ in a Hilbert space $\mathcal{H}$ gives rise to a linear operator $\widehat{O}_{\vec{u},\vec{v}}$ defined by

$$\widehat{O}_{\vec{u},\vec{v}}\,\vec{\psi} := \langle \vec{v}, \vec{\psi} \rangle\, \vec{u}. \tag{3.22}$$

This useful construction is especially easy to handle using Dirac notation, where $\widehat{O}_{\vec{u},\vec{v}}$ is written as $|u\rangle\langle v|$ with

$$(\,|u\rangle\langle v|\,)\,|\psi\rangle := |u\rangle\langle v|\psi\rangle. \tag{3.22$'$}$$

In particular, if $\{\,|e_1\rangle, |e_2\rangle, \ldots, |e_N\rangle\}$ is an orthonormal basis set for the Hilbert space $\mathcal{H}$, then $(\,|e_i\rangle\langle e_i|\,)\psi = |e_i\rangle\langle e_i|\psi\rangle$, and (2.63′) can be written as

$$\widehat{\mathbb{1}} = \sum_{i=1}^{N} |e_i\rangle\langle e_i|\,. \tag{3.23}$$

This is known as the *resolution of the identity* associated with the given basis set. We shall return shortly to this idea in the context of the spectral theorem for self-adjoint operators.

Operators like $|u\rangle\langle v|$ can also be thought of as acting on bras according to the rule $\langle\phi|\,(\,|u\rangle\langle v|\,) := \langle\phi|u\rangle\langle v|\,.$ This is consistent with the relation [Exercise!]

$$(\,|u\rangle\langle v|\,)^{\dagger} = |v\rangle\langle u|\,. \tag{3.24}$$

Thus, for example, the expression

$$\langle\phi|\psi\rangle = \sum_{i=1}^{N}\langle e_i|\phi\rangle^{*}\langle e_i|\psi\rangle = \sum_{i=1}^{N}\langle\phi|e_i\rangle\langle e_i|\psi\rangle \tag{3.25}$$

can be viewed as coming from the identity $\langle\phi|\psi\rangle \equiv \langle\phi|\,\widehat{\mathbb{1}}\,|\psi\rangle$, with the unit operator $\widehat{\mathbb{1}}$ replaced by the right hand side of Eq. (3.23).

It is a straightforward exercise to prove the following.

1. The results in Eq. (3.20) show that the matrix representative of the adjoint of an operator with respect to an orthonormal basis satisfies

$$(A^{\dagger})_{ij} = A^{*}_{ji} \quad i,j = 1,2,\ldots,N. \tag{3.26}$$

Note that if $\widehat{A}$ is a self-adjoint operator then $A_{ij} = A^{*}_{ji}$, *i.e.*, its matrix representative is a hermitian matrix in the usual sense.

2. The adjoints of operators obey the following equations

$$(\widehat{A}\widehat{B})^\dagger = \widehat{B}^\dagger\widehat{A}^\dagger \tag{3.27}$$

$$(\lambda\widehat{A})^\dagger = \lambda^*\widehat{A}^\dagger \text{ for all } \lambda \in \mathbb{C}, \tag{3.28}$$

$$(\widehat{A}^\dagger)^\dagger = \widehat{A}. \tag{3.29}$$

3. The matrix corresponding to the product $\widehat{A}\widehat{B}$ of two operators $\widehat{A}$ and $\widehat{B}$ is equal to the *matrix* product of the matrices representing $\widehat{A}$ and $\widehat{B}$. This is another way of understanding why the product of two matrices is defined as it is.

3.2 Projection Operators

Projection operators are operators that project a vector onto some subspace of the Hilbert space. As we shall see, they are self-adjoint and have 0 and 1 as their only eigenvalues. They correspond therefore to 'binary-valued' observables and can be understood as propositions about properties of the quantum system. As such, they play a central role in discussions of the conceptual foundations of quantum theory (in particular, see the short discussion of quantum logic in Section 9.2).

First, some preliminary definitions:

Definition

1. A linear subspace W of a Hilbert space $\mathcal{H}$ is topologically *closed* if for every strongly convergent sequence of vectors $\vec{v}_1, \vec{v}_2, \ldots$ lying in W, the limit vector $\vec{v} \in \mathcal{H}$ also belongs to W. This is true of every linear subspace W if the dimension of $\mathcal{H}$ is finite, but it is easy to find non-closed linear subspaces of an infinite-dimensional Hilbert space[3].

   For any linear subspace W of $\mathcal{H}$ there exists a smallest linear subspace of $\mathcal{H}$ that is closed and that contains W as a subspace. This is known as the *closure* of W, and is denoted $\overline{W}$.

[3] For example, the set of all *finite* linear combinations of an orthonormal basis set $\{\vec{e}_1, \vec{e}_2, \ldots\}$ of $\mathcal{H}$ is such a subspace.

2. Two linear subspaces W_1 and W_2 of $\mathcal{H}$ are *orthogonal* if every vector in W_1 is orthogonal to every vector in W_2. The *orthogonal complement* $W^\perp$ of a linear subspace W of $\mathcal{H}$ is the set of all vectors that are orthogonal to every vector in[4] W:

$$W^\perp := \{\vec{\psi} \in \mathcal{H} \mid \text{for all } \vec{w} \in W, \langle \vec{w}, \vec{\psi} \rangle = 0\}. \tag{3.30}$$

Clearly W and $W^\perp$ are orthogonal subspaces of $\mathcal{H}$.

It is easy to see that $W^\perp$ is a linear subspace of $\mathcal{H}$. It is also topologically closed, although we shall not prove that here (it is trivially true in finite dimensions since every linear subspace is topologically closed). It is interesting to note that if $\mathcal{H}$ has a finite dimension then $(W^\perp)^\perp = W$, whereas if the dimension of $\mathcal{H}$ is infinite then $(W^\perp)^\perp$ is the closure $\overline{W}$ of W.

The main geometrical idea in the present section is that, given a topologically closed subspace W of $\mathcal{H}$, any vector $\vec{\psi}$ can be decomposed as a unique sum

$$\vec{\psi} = \vec{\psi}_W + \vec{\psi}_{W^\perp} \tag{3.31}$$

of vectors $\vec{\psi}_W$ and $\vec{\psi}_{W^\perp}$ that lie in W and $W^\perp$ respectively. This is illustrated heuristically in Figure 3.1. It is easy to show that any such decomposition is unique [Exercise!]. The precise mathematical definition of $\vec{\psi}_W$ and $\vec{\psi}_{W^\perp}$ is straightforward. Let $\{\vec{f}_1, \vec{f}_2, \ldots\}$ be any orthonormal basis for the subspace W. Then define[5]

$$\vec{\psi}_W := \sum_i \langle \vec{f}_i, \vec{\psi} \rangle \, \vec{f}_i \tag{3.32}$$

and

$$\vec{\psi}_{W^\perp} := \vec{\psi} - \vec{\psi}_W. \tag{3.33}$$

It is an easy (but important!) exercise to show that the vectors $\vec{\psi}_W$ and $\vec{\psi}_{W^\perp}$ thus defined do not depend on the choice of orthonormal basis $\{\vec{f}_1, \vec{f}_2, \ldots\}$ for the subspace W.

The map $\vec{\psi} \mapsto \vec{\psi}_W$ is clearly linear, and hence can be regarded as the action of some operator $\widehat{P}_W$; *i.e.*,

$$\widehat{P}_W \vec{\psi} := \vec{\psi}_W. \tag{3.34}$$

---

[4]The notation $\{x \in X \mid P(x)\}$ means the subset of all elements x of the set X for which the proposition $P(x)$ is true.

[5]Strictly speaking, if $\mathcal{H}$ has infinite dimension, it is necessary to show that the sum in Eq. (3.32) converges strongly, but we shall not bother with such niceties here.

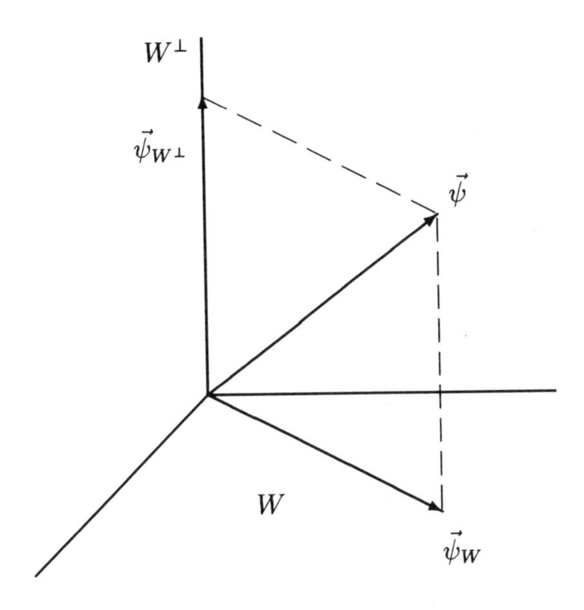

Figure 3.1: The decomposition of the vector $\vec{\psi}$

This is called the *projection operator* onto the subspace W. The projector onto the orthogonal complement $W^\perp$ is $\widehat{P}_{W^\perp}$, and is equal to $\widehat{\mathbb{1}} - \widehat{P}_W$.

Comments

1. The simplest type of subspace of $\mathcal{H}$ is the one-dimensional space of all complex multiples of some unit length vector $\vec{u}$. The projector onto this subspace is the operator $\widehat{O}_{\vec{u},\vec{u}}$ (see Eq. (3.22)); in Dirac notation it is just the 'butterfly operator' $|u\rangle\langle u|$.

2. If $\vec{\psi} \in W$ then $\widehat{P}_W\vec{\psi} = \vec{\psi}$, and if $\vec{\psi} \in W^\perp$ then $\widehat{P}_W\vec{\psi} = 0$. Conversely, if $\vec{\psi} \in V$ is such that $\widehat{P}_W\vec{\psi} = \vec{\psi}$, then $\vec{\psi}$ necessarily belongs to the subspace W; similarly $\widehat{P}_W\vec{\psi} = 0$ implies $\vec{\psi} \in W^\perp$.

More generally, it is straightforward to show that a non-trivial projection operator $\widehat{P}_W$ has just two eigenvalues 1 and 0. Both eigenvalues are degenerate (assuming the dimensions of W and $W^\perp$ are larger than one) and the corresponding eigenspaces are W and $W^\perp$ respectively. Note that if W is an infinite-dimensional subspace of $\mathcal{H}$, then $\widehat{P}_W$ is an example of an operator with an infinitely-degenerate eigenvalue.

3. A projection operator $\widehat{P}_W$ satisfies

$$\widehat{P}_W^2 = \widehat{P}_W \tag{3.35}$$
$$\widehat{P}_W^\dagger = \widehat{P}_W. \tag{3.36}$$

Conversely, let $\widehat{P}$ be any operator that satisfies the two conditions

$$\widehat{P}^2 = \widehat{P} \tag{3.37}$$
$$\widehat{P}^\dagger = \widehat{P}. \tag{3.38}$$

Then $\widehat{P}$ is a projection operator, and the subspace of $\mathcal{H}$ onto which it projects is $\mathcal{H}_{\widehat{P}} := \{\vec{\psi} \in \mathcal{H} \mid \widehat{P}\vec{\psi} = \vec{\psi}\}$. Indeed, Eq. (3.37–3.38) are often used to *define*[6] what is meant by a projection operator in an algebraic way.

4. Projection operators $\widehat{P}_1$ and $\widehat{P}_2$ are said to be *orthogonal* if $\widehat{P}_1\widehat{P}_2 = 0 = \widehat{P}_2\widehat{P}_1$. This is equivalent to the statement that the subspaces onto which they project are orthogonal. For example, if $\{\,|e_1\rangle, |e_2\rangle, \ldots\}$ is an orthonormal basis set for $\mathcal{H}$, the projection operators $|e_i\rangle\langle e_i|$ and $|e_j\rangle\langle e_j|$ are orthogonal if $i \neq j$.

In general, commuting operators correspond to observables that can be 'measured simultaneously' (see section 6.3). Thus commuting projectors are associated with propositions that can be simultaneously asserted in some sense.

Note that if $\widehat{P}_1$ and $\widehat{P}_2$ are orthogonal, then $\widehat{P}_1 + \widehat{P}_2$ is also a projector. Indeed, this operator is clearly self-adjoint, and

$$(\widehat{P}_1 + \widehat{P}_2)(\widehat{P}_1 + \widehat{P}_2) = \widehat{P}_1^2 + \widehat{P}_2^2 + \widehat{P}_1\widehat{P}_2 + \widehat{P}_2\widehat{P}_1 = \widehat{P}_1 + \widehat{P}_2 \tag{3.39}$$

since $\widehat{P}_1\widehat{P}_2 = 0 = \widehat{P}_2\widehat{P}_1$.

5. The notion of decomposing a vector into components that lie in a subspace W and its orthogonal complement $W^\perp$, as in Eq. (3.31), can be generalised in a significant way. Thus, let $\{W_1, W_2, \ldots, W_M\}$ be any

[6]Sometimes a projection operator is defined to be any operator satisfying Eq. (3.37) only; addition of the hermiticity condition Eq. (3.38) then gives rise to an *orthogonal* projection operator. However, I shall use the term projection operator only if both Eq. (3.37) and Eq. (3.38) are satisfied.

collection of closed subspaces of $\mathcal{H}$ (M may be infinite). Then $\mathcal{H}$ is said to be the *direct sum* of these subspaces if (i) they are mutually orthogonal, and (ii) each vector in $\mathcal{H}$ is equal to the sum of its projections onto the subspaces; *i.e.*, for any vector $\vec{\psi} \in \mathcal{H}$ we have

$$\vec{\psi} = \widehat{P}_{W_1}\vec{\psi} + \widehat{P}_{W_2}\vec{\psi} + \cdots + \widehat{P}_{W_M}\vec{\psi}. \tag{3.40}$$

We write

$$\mathcal{H} = W_1 \bigoplus W_2 \bigoplus \cdots \bigoplus W_M, \tag{3.41}$$

or $\mathcal{H} = \bigoplus_{k=1}^{M} W_k$.

The corresponding projection operators $\widehat{P}_{W_1}, \widehat{P}_{W_2}, \ldots, \widehat{P}_{W_M}$ are pairwise orthogonal

$$\widehat{P}_{W_i}\widehat{P}_{W_j} = \delta_{ij}\widehat{P}_{W_i}, \; i,j = 1,2,\ldots,M \tag{3.42}$$

and, from Eq. (3.40), satisfy

$$\widehat{P}_{W_1} + \widehat{P}_{W_2} + \cdots + \widehat{P}_{W_M} = \widehat{\mathbb{1}}. \tag{3.43}$$

This is called a *resolution of the identity*.

For example, if W is any topologically closed subspace of $\mathcal{H}$, then $\mathcal{H} = W \oplus W^{\perp}$, and the analogue of Eq. (3.43) is the identity $\widehat{\mathbb{1}} = \widehat{P}_W + (\widehat{\mathbb{1}} - P_W)$. Generalised decompositions of the Hilbert space of the form Eq. (3.41) (and the associated equation Eq. (3.43)) are an integral part of the mathematical formalism of quantum theory.

Note that a special case is when W_i is the one-dimensional subspace spanned by an element $\vec{e}_i$ of an orthonormal basis set $\{\vec{e}_1, \vec{e}_2, \ldots, \vec{e}_M\}$ of $\mathcal{H}$. Eq. (3.40) is then just the usual expression for the expansion of $\vec{\psi}$ with respect to the basis set.

3.3 Spectral Theory for Self-Adjoint Operators

From a mathematical perspective, self-adjoint differential operators play such a central role in wave-mechanics because (i) their eigenvalues are real numbers (and can therefore represent the results of physical measurements); (ii) eigenfunctions corresponding to different eigenvalues have vanishing overlap functions; and (iii) any wave function can be expanded as a linear combination of the eigenfunctions of such an operator. It is of major importance that there are precise analogues of these results for *any* Hilbert space.

The first two are rather elementary, and are contained in the following well-known theorem.

Theorem

1. The eigenvalues of a self-adjoint operator $\widehat{A}$ are real numbers.
2. Eigenvectors corresponding to two different eigenvalues are orthogonal.

Proof

1. Let $\widehat{A}\vec{u} = a\vec{u}$. Then

$$\langle \widehat{A}\vec{u}, \vec{u}\rangle = \langle a\vec{u}, \vec{u}\rangle = a^*\langle \vec{u}, \vec{u}\rangle. \tag{3.44}$$

But the operator $\widehat{A}$ is self-adjoint, and so

$$\langle \widehat{A}\vec{u}, \vec{u}\rangle = \langle u, \widehat{A}\vec{u}\rangle = \langle \vec{u}, a\vec{u}\rangle = a\langle \vec{u}, \vec{u}\rangle. \tag{3.45}$$

Subtracting the first equation from the second we get $(a - a^*)\langle \vec{u}, \vec{u}\rangle = 0$. But, since $\vec{u} \neq 0$, we have $\langle \vec{u}, \vec{u}\rangle \neq 0$ and hence $a = a^*$, which proves that a is real.

2. Let $\widehat{A}\vec{u}_1 = a_1\vec{u}_1$ and $\widehat{A}\vec{u}_2 = a_2\vec{u}_2$. Taking the scalar product of the first equation with $\vec{u}_2$ gives

$$\langle \vec{u}_2, \widehat{A}\vec{u}_1\rangle = \langle \vec{u}_2, a_1\vec{u}_1\rangle = a_1\langle \vec{u}_2, \vec{u}_1\rangle. \tag{3.46}$$

But, since $\widehat{A}$ is self-adjoint, the left hand side is

$$\langle \widehat{A}\vec{u}_2, \vec{u}_1\rangle = \langle a_2\vec{u}_2, \vec{u}_1\rangle = a_2^*\langle \vec{u}_2, \vec{u}_1\rangle = a_2\langle \vec{u}_2, \vec{u}_1\rangle, \tag{3.47}$$

where we have used $a_2^* = a_2$. Hence $(a_1 - a_2)\langle \vec{u}_2, \vec{u}_1\rangle = 0$. But $a_1 \neq a_2$, and so

$$\langle \vec{u}_2, \vec{u}_1\rangle = 0 \tag{3.48}$$

as claimed. **QED**

Finally we come to the so-called 'spectral theorem': a result upon which hangs the entire edifice of quantum theory. This is the fact that any vector can be expanded uniquely as a linear combination of eigenvectors of a self-adjoint operator.

To make this statement precise it is necessary to return to the notion of degenerate eigenvalues. If a_m is a $d(m)$-fold degenerate eigenvalue of $\widehat{A}$ with linearly independent eigenvectors $\vec{u}_{m1}, \vec{u}_{m2}, \ldots, \vec{u}_{m\,d(m)}$ then, as discussed in the context of Eq. (3.10), the set of eigenvectors with eigenvalue a_m is a linear subspace of $\mathcal{H}$ with dimension $d(m)$. It is clear that the set $\{\vec{u}_{m1}, \vec{u}_{m2}, \ldots, \vec{u}_{m\,d(m)}\}$ of vectors in this eigenspace can always be *chosen* to be orthonormal, so that

$$\langle \vec{u}_{mi}, \vec{u}_{mj} \rangle = \delta_{ij} \quad i, j = 1, 2, \ldots, d(m). \tag{3.49}$$

If this is done for each eigenvalue a_m of $\widehat{A}$ we get a collection of vectors $\{\vec{u}_{mi},\ m = 1, 2, \ldots, M,\ i = 1, \ldots, d(m)\}$, where $a_1, a_2, \ldots, a_M$ are the different eigenvalues of the operator (in the infinite-dimensional case M could also be infinite). It follows from the theorem above that this entire set of vectors is orthonormal

$$\langle \vec{u}_{mi}, \vec{u}_{nj} \rangle = \delta_{mn}\delta_{ij} \tag{3.50}$$

where $n, m = 1, 2, \ldots, M$, $i = 1, 2, \ldots, d(m)$ and $j = 1, 2, \ldots, d(n)$.

The fundamental *spectral theorem* for a finite-dimensional Hilbert space is as follows.

Theorem

The set of all eigenvectors of a self-adjoint operator $\widehat{A}$ is an orthonormal basis set for $\mathcal{H}$. Thus, in terms of the notation above, any vector $\vec{\psi} \in \mathcal{H}$ can be expanded as

$$\boxed{\vec{\psi} = \sum_{m=1}^{M} \sum_{j=1}^{d(m)} \psi_{mj} \vec{u}_{mj}} \tag{3.51}$$

where the expansion coefficients $\psi_{mj} \in \mathbb{C}$ are given explicitly by

$$\psi_{mj} = \langle \vec{u}_{mj}, \vec{\psi} \rangle. \tag{3.52}$$

Comments

1. This is a famous result in linear algebra, but it would be too much of a diversion to prove it here. The situation for an infinite-dimensional Hilbert space is far more complicated but, broadly speaking, the theorem applies as it stands to any 'well-behaved' self-adjoint operator whose eigenvalues form a discrete subset of the real numbers. A good example is the Hamiltonian of the simple harmonic oscillator whose eigenvalues are $(n + \frac{1}{2})\hbar\omega$ with $n = 0, 1, 2, \ldots$.

2. Quantum theory also uses operators with a *continuous* spectrum: for example, the momentum operator $-i\frac{d}{dx}$ has eigenfunctions $u_k(x) := e^{ikx}$, with the eigenvalue k ranging over the entire real line. This suggests that the sums in the statement of the spectral theorem should be replaced by integrals, and we shall discuss this briefly in Section 6.2. But note that, strictly speaking, these 'eigenvectors' are not *in* the appropriate Hilbert space $L^2(\mathbb{R})$ at all since they have an infinite norm! This property is characteristic of operators with a continuous spectrum, and is yet another example of the problems that arise in attempts to handle infinite-dimensional spaces. Fortunately, for most purposes one can avoid using such operators, and we shall do so wherever possible.

3. In Dirac notation, the basic expansion theorem reads

$$|\psi\rangle = \sum_{m=1}^{M} \sum_{j=1}^{d(m)} \langle a_m, j|\psi\rangle \, |a_m, j\rangle, \tag{3.53}$$

with an associated resolution of the identity

$$\widehat{\mathbb{1}} = \sum_{m=1}^{M} \sum_{j=1}^{d(m)} |a_m, j\rangle\langle a_m, j| \, . \tag{3.54}$$

The operator $\widehat{A}$ itself can be written in the important form [Exercise!]

$$\widehat{A} = \sum_{m=1}^{M} \sum_{j=1}^{d(m)} a_m \, |a_m, j\rangle\langle a_m, j| \tag{3.55}$$

which is known as the *spectral representation* of $\widehat{A}$.

4. The operator

$$\widehat{P}_m := \sum_{j=1}^{d(m)} |a_m, j\rangle\langle a_m, j| \tag{3.56}$$

is the projection operator onto the subspace of eigenvectors of $\widehat{A}$ with eigenvalue a_m; we shall refer to it as a *spectral projector*. If it is necessary to emphasise the value of the eigenvalue, the operator $\widehat{P}_m$ will be written as $\widehat{P}_{A=a_m}$. This projector is independent of the set of pairwise-orthogonal vectors $\{\, |a_m, 1\rangle,\ |a_m, 2\rangle, \ldots,\ |a_m, d(m)\rangle\}$ chosen to span the eigenspace.

The projectors $\widehat{P}_m$ are pairwise orthogonal

$$\widehat{P}_m \widehat{P}_n = \delta_{mn} \widehat{P}_m, \tag{3.57}$$

and can be used to write the basic expansion theorem Eq. (3.53) as

$$|\psi\rangle = \sum_{m=1}^{M} \widehat{P}_m\, |\psi\rangle, \tag{3.58}$$

which corresponds to the resolution of the identity

$$\widehat{1} = \sum_{m=1}^{M} \widehat{P}_m. \tag{3.59}$$

This shows that $\mathcal{H}$ is the direct sum of the eigenspaces $\mathcal{H}_{\widehat{P}_1}, \mathcal{H}_{\widehat{P}_2}, \ldots, \mathcal{H}_{\widehat{P}_M}$. Note that the expressions Eq. (3.58–3.59) are manifestly independent of the choice of basis vectors in the various eigenspaces.

The spectral representation Eq. (3.55) can be rewritten in terms of these spectral projectors as

$$\boxed{\widehat{A} = \sum_{m=1}^{M} a_m\, \widehat{P}_m} \tag{3.60}$$

This result is one of the cornerstones of the mathematical framework of modern quantum theory. We shall return to it frequently in what follows.

5. If Δ is some subset of the real line it is meaningful to ask for the projector $\widehat{P}_{A\in\Delta}$ onto the subspace of all eigenvectors whose eigenvalues lie in $\Delta \subset \mathbb{R}$. It is easy to see from the orthogonality condition Eq. (3.57) that this is just

$$\widehat{P}_{A\in\Delta} = \sum_{a\in\Delta} \widehat{P}_{A=a}. \tag{3.61}$$

This expression plays an important role in the general analysis of quantum theory.

6. As was emphasised earlier, a given operator $\widehat{A}$ on a vector space has many matrix representations, corresponding to different choices for the set of basis vectors used in the construction of the matrix from the operator. A particularly useful set for a self-adjoint operator is the collection of its eigenvectors. For example, if $\widehat{A}$ is multiplicity-free, the matrix elements with respect to the basis set $\{\,|a_1\rangle,\ |a_2\rangle, \ldots,\ |a_M\rangle\}$ are

$$A_{ij} := \langle a_i|\,\widehat{A}\,|a_j\rangle = a_i\langle a_i|a_j\rangle = a_i\delta_{ij}. \tag{3.62}$$

Hence $\widehat{A}$ is represented by the diagonal matrix

$$\begin{pmatrix} a_1 & 0 & \ldots & 0 \\ 0 & a_2 & \ldots & 0 \\ \vdots & \vdots & \ddots & \vdots \\ 0 & 0 & \ldots & a_M \end{pmatrix}. \tag{3.63}$$

A simple example of such a representation is afforded by the Pauli spin matrix σ_z in Eq. (1.23). Thus, in writing the Pauli spin matrices in the form Eq. (1.23), it has been assumed implicitly that the basis set of $\mathbb{C}^2$ to be used is that given by the eigenvectors $\binom{1}{0}$ and $\binom{0}{1}$ of σ_z.

Chapter 4

PROPERTIES IN CLASSICAL PHYSICS

4.1 What is a Thing?

An exposition of any area of physics will inevitably contain terms that form part of the general scientific background of the age and culture within which they are employed. The meaningfulness and applicability of such terms is usually deemed to be 'obvious', and therefore not worthy of further explication. But from time to time new concepts arise that challenge this pre-established order of truths and necessitate a radical reappraisal of the foundations of the subject. In twentieth-century physics, the two major examples of such a paradigm shift are the theory of relativity and quantum theory. The former caused a major reassessment of the concepts of space and time; the latter challenges our ideas of existence itself.

Examples of seemingly innocuous terms that arise in discussions of quantum theory include 'system', 'observable', 'property', 'physical quantity', 'measurement', 'state', 'causality', 'determinism'. Any serious discussion of such concepts leads at once into deep issues in the general philosophy of science, such as:

- the object matter of scientific investigations;
- the status of theoretical terms and postulated entities;

- the relation of mathematical models to the physical world;
- the question of if, and how, scientific statements can be verified or falsified;
- the nature of space and time;
- the meaning of probability.

Issues of this type lead us to augment our list of 'innocuous' terms with problematic words such as 'realism', 'instrumentalism', 'operationalism', 'empiricism', 'reductionism', and the like. Properly speaking, these all need careful consideration.

Such a general investigation would be far beyond the intended scope of these notes. Nevertheless, something must be said in order that the revolutionary nature of quantum theory can be appreciated fully. In classical physics, the interpretation of the 'innocuous' terms is broadly in line with the 'commonsense' view of reality, and therefore their meaning might indeed be regarded as being obvious and relatively uncontentious. However, in quantum theory the situation is radically altered, and the cumulative shift in meaning of these basic ideas has tended towards a view of reality that is profoundly different. To facilitate understanding this fundamental change it is helpful to start with a short exposition of what the basic conceptual position of classical physics really is. Of particular interest here is the way in which the specific mathematical tools used to describe a physical system can encode a quite definite philosophical position towards the physical world.

Let us start with the word *system*, which simply refers to the object whose properties—typically the values of physical quantities like energy, charge, *etc.*—are being studied. But what is meant by an 'object'? And what is a 'property'?; or a 'physical quantity'? Is the significance of these terms really so obvious? We are confronted here with the question whose seeming innocence is so singularly deceptive:

What is a 'thing'?

This splendid question launched Western philosophy two and a half thousand years ago and, together with the complementary query "what is the nature of our *knowledge* of things?", continues to haunt us today.

The matter is certainly not trivial: at least, it is not like asking what is a 'cat', or 'a theoretical physicist', or even 'a nice cup of tea'. One might attempt to answer questions of that type by specifying a list of properties that characterise the entity concerned, although even this is notoriously difficult. But the question "what is a thing?" cannot be read in this way. It does not ask for the properties of some particular thing that would distinguish it from any other, but rather for the essence of things in general; what Heidegger[1] memorably called the 'thingness of things'. In particular, the question does not admit an answer of the form "A thing is a ...", since that little word 'is' is part of the problem. To ask "what is a thing?" invites the bigger question "what does it mean 'to be'?". If posed in the context of physics, this leads inevitably to the great debate between instrumentalists and realists on the nature of the scientific enterprise in general, especially the role and status of mathematical descriptions of the world.

It is striking that, of all the modern sciences, only quantum physics seems to have been obliged to face the issue of Being directly. However, in truth, no branch of scientific study can avoid this central question entirely; indeed, it should lie at the heart of *any* attempt to grasp the nature of reality, be it scientific or otherwise. Unfortunately, most people never address the problem, or even realise that there is one to be addressed. Instead, the answer to the unasked question is assumed as an *a priori* feature of reality, and is thereby imposed on the world as part of the 'obvious' truths of the discipline concerned. In a notable quote by the physicist Eddington (1920):

> "We have found that where science has progressed the farthest, the mind has but regained from nature that which the mind has put into nature. We have found a strange foot-print on the shores of the unknown. We have devised profound theories, one after another, to account for its origin. At last, we have succeeded in reconstructing the creature that made the foot-print. And Lo! it is our own."

[1]For example, see the work (with strong Kantian overtones) by Heidegger (1967).

Or, as the Argentinian writer Borges (1964) put it,

> "A man sets himself the task of portraying the world. Through the years he peoples a space with images of provinces, kingdoms, mountains, bays, ships, islands, fishes, rooms, instruments, stars, horses, and people. Shortly before his death, he discovers that the patient labyrinth of lines traces the image of his face."

In fact, the 'truths' concerned are very far from being obvious and, as these authors were trying to emphasise, can be as much a reflection of our own projections as they are of reality itself.

In this context, one should not forget that an attachment to a particular philosophical position can have powerful emotional overtones. The interpretation of quantum theory is a potent example of this phenomenon: it is not unusual to find a physicist, or philosopher of science, defending a specific position with a fervour and passion that far outreaches the degree of emotion normally associated with scientific beliefs; indeed, sometimes it is as if his or her very existence depended on the outcome of the debate.

The emotional character of archetypal projections was frequently emphasised by C.G. Jung in his studies of the unconscious world. For example, in relation to the concept of 'causality' (reflecting, perhaps, the earlier strictures of Hume) he wrote (Jung 1983)

> "Another inexhaustible source of happiness can be the gratification of the causal instinct."

and

> "Just as the sexual drive frequently transforms man into a monster, so the elementary category of causality can assume the character of a need, an insatiable craving which overruns everything, and which people will even sacrifice their lives to gratify. It is an indefatigable longing which inflames us, which makes us despise all the works and ordinances of man, which makes us smile when others are weeping."

Of course, it could be argued that the very essence of the scientific method is to force Nature to respond to our investigations using certain pre-assigned categories. However, the subject for concern is not scientific methodology *per se* but rather the problems that arise when the projections are made unconsciously, so that we do not appreciate the extent to which what we call 'reality' may be created, rather than discovered. These cautionary remarks should be kept in mind when assessing the peristalithic debate on the meaning of quantum theory.

4.2 The State Space of Classical Physics

4.2.1 General Concepts

A commonsense answer to "what is a thing?" might focus on three types of entity. The first is 'objects': things that are near-at-hand (or become so after using a microscope or telescope), and are presumably the main concern of the physical sciences. However, we also talk about things as states of affairs—we ask "how are things with you?", or respond "things are not going well at the moment"—and, as an extreme use of the word 'thing', we can mean anything that is not nothing; in this sense, the number seven is a thing, as is 'blueness', 'beauty', 'a vice-chancellor' *etc.* In what follows, we shall be concerned only with the first of these classes—the object-things—and can therefore restrict our investigation of the thingness of things to this special case.

What then is a 'thing' (or 'physical system') in this special physical sense? The approach of classical physics takes for granted the 'thing-in-itself' and concentrates on the bundle of *properties*, or attributes, that adhere to the thing and make it what it is. In this context, a 'property' refers to the values of some specified class of physical quantities lying in specified ranges. Thus we talk about the mass of an electron, its electric charge, its position and its velocity; or we talk of the position of the pencil on a desk, its length, and the direction in which it points.

Note that mass or charge are examples of *internal* physical quantities—those whose values refer to the constitution of the thing itself—whereas position, direction, velocity, are *external* physical quantities whose numerical values specify how the object appears in, or relates to, the framework of space and time. In practice, much of classical physics is divided into the study of *particles*—associated with bundles of properties that are spatially localised; and *fields*—associated with properties that vary smoothly across the space-time continuum[2]. In the latter case, the distinction between internal and external properties becomes rather blurred since the properties of a field are tied intimately to the structure of space and time.

The notion of an object as the bearer of determinative properties lies at the heart of realism, which is one of the main Western, 'commonsense' philosophical positions. Indeed, for the purposes of this book, the phrase "realist interpretation" will primarily *mean* one in which (i) it is deemed appropriate to talk about physical quantities 'having' values at any time; and (ii) propositions about such possessed values can be handled using the tools of conventional propositional logic[3]. In this context, note that the idea of an object bearing properties is encoded in the structure of our language itself: the subject-predicate form of simple sentences (for example, "the desk is two meters wide") reflects the idea of properties adhering to things, and thereby our comonsense understanding of the question "what is a thing?".

The epistemological[4] question of how we can have knowledge of the properties of an object is answered in physics by the notion of *measurement*, *i.e.*, any physical operation by which the value of a physical quantity can be determined (perhaps only to a certain accuracy) and recorded. Such a picture is in accord with the general object-subject split of scientific methodology whereby part of the natural world is deliberately isolated from its environment so that theoretical and experimental investigations may proceed unhampered by any influence from the rest of the universe.

From the perspective of classical physics, the separation of observer and system has no fundamental significance. Observer and observed are both parts of a single, objectively-existing world in which, ontologically speaking, both have equal status, and are potentially describable by the

---

[2]One might wish to add continuum mechanics as a separate branch of classical physics.

[3]Whether or not this is, in fact, possible in quantum theory is one of the major issues to be discussed.

[4]Epistemology is the study of knowledge, of what we can know about what exists. The complementary concept is *ontology*: the study of being, of what can be said in general about what exists.

same physical laws. Similarly, there is nothing special about the concepts of 'measurement' or 'observable'. The reason why a measurement of an observable quantity yields one value rather than another is simply because the quantity *has* that value at the time the measurement is made. Thus properties are intrinsically attached to the object as it exists in the world, and measurement is nothing more than a particular type of physical interaction designed to display the value of a specific quantity. Of course, this presupposes that 'perfect' measurements exist which have this desirable property of revealing what is actually the case; such an assumption is not totally trivial.

The general view sketched above of the nature of 'things' also determines how we commonly understand the significance of time. We say that things "change in time", by which it is meant that some, or all, of the internal or external properties of an object are time-dependent. Furthermore, in classical physics it is assumed that these changes are *deterministic*, *i.e.*, a knowledge of a sufficiently complete set of properties of an object at one time suffices to predict with certainty the properties at any later (or earlier) time.

From one perspective, the central challenge for a theoretical physicist is to find the types of mathematical structure that can successfully encode this, essentially philosophical, view of the significance and structure of the properties of an object. In particular, we must understand how the logical operations instinctively used to handle propositions about such things are related to, or are consistent with, the mathematical framework employed in describing the system.

In practice, the concepts of physical quantities and properties are coded scientifically into the notion of the space $\mathcal{S}$ of *states* of a system, with the understanding that, at any given time, a unique member of $\mathcal{S}$ can be assigned to the system. Much of the underlying philosophical position of a theory is reflected in the precise role played by its states. In the case of classical physics, the state assignment is supposed to satisfy the following conditions:

S1 A specification of the state at any time suffices to determine the values[5]

[5]This could be put in more operational terms by saying that a specification of the state at any time suffices to predict with certainty the numerical results of all possible measurements of observable quantities that could be made at that time.

at that time of all physical quantities pertaining to the system.

S2 The state at any time t_2 is determined[6] uniquely by the state at any earlier time t_1. This *principle of causality* is how strict determinism finds its way into physics.

For many types of system (for example, Newtonian point particles; the classical electromagnetic field) the state s at a time t_1 determines not only the state at a later time $t_2 > t_1$ but also the state at any *earlier* time $t_0 < t_1$, *i.e.*, the state that evolved into s in the time duration $t_1 - t_0$.

The mathematical implication of the first requirement is that to each physical quantity A of a system (for example, velocity of this particle, position of that one, total energy) there corresponds a function $f_A : \mathcal{S} \to \mathbb{R}$ such that $f_A(s)$ is the value which A possesses when the state is s.

The second requirement implies the existence of a two-parameter family of 'dynamical' maps $T_{t_2t_1} : \mathcal{S} \to \mathcal{S}$ such that if the state of the system at time t_1 is $s \in \mathcal{S}$, the state at some (later) time t_2 will be $T_{t_2t_1}(s)$. The maps $T_{t_2t_1}$ clearly satisfy the conditions[7]

$$T_{tt} = \mathrm{id} \quad \text{for all } t \in \mathbb{R} \tag{4.1}$$

and

$$T_{t_3t_2}T_{t_2t_1} = T_{t_3t_1} \tag{4.2}$$

if $t_1 \leq t_2 \leq t_3$.

A typical question that can be posed meaningfully in this context, and one that is of direct physical interest, is: "If, at time $t = t_1$ a physical quantity A has the value a, what will be its value at some later time $t = t_2$?". This question can only be answered if the state in which A has the value a is *unique*, *i.e.*, there exists $s \in \mathcal{S}$ such that[8] $f_A^{-1}(\{a\}) = \{s\}$. In

[6]Recent developments in the theory of 'chaotic' motion have enforced important qualifications in the way classical determinism is viewed. However, this does not affect the line of argument being developed here.

[7]The notation 'id' refers to the identity map from $\mathcal{S}$ to itself.

[8]In general, if f is a map from a set X to a set Y, and if W is a subset of Y, then $f^{-1}(W)$ is defined to be the subset of X of all elements that are mapped by f into W, *i.e.*, $f^{-1}(W) := \{x \in X \mid f(x) \in W\}$. This 'set-inverse' operation should not be confused with the inverse function $f^{-1} : Y \to X$, which only exists if f is a bijection. Note that $\{a\}$ denotes the subset of $\mathbb{R}$ consisting of the single element $a \in \mathbb{R}$.

this case the value of A at time t_2 is just $f_A(T_{t_2t_1}(s))$. However, in general the subset $f_A^{-1}(\{a\})$ of $\mathcal{S}$ will have more than one element, in which case nothing further can be said.

The question can be posed more generally for a *collection* $A_1, A_2, \ldots, A_n$ of quantities whose initial values are $a_1, a_2, \ldots, a_n$ respectively. We shall say that this collection of properties *fixes* a state $s \in \mathcal{S}$ if

$$f_{A_1}^{-1}(\{a_1\}) \bigcap f_{A_2}^{-1}(\{a_2\}) \bigcap \cdots \bigcap f_{A_n}^{-1}(\{a_n\}) = \{s\} \tag{4.3}$$

i.e., if there exists a unique state $s \in \mathcal{S}$ with the given properties. In this case, the value of any of the quantities A_i at time t_2 is given by $f_{A_i}(T_{t_2t_1}(s))$.

4.2.2 The Example of a Point Particle

To illustrate the idea of states and their mathematical representation, consider a point particle of mass m moving in one dimension under the influence of a force $F(x, \frac{dx}{dt})$ (if present, a velocity-dependent term typically corresponds to some type of frictional force). The motion of the system is described by Newton's second law

$$m\frac{d^2x(t)}{dt^2} = F(x(t), \frac{dx(t)}{dt}) \tag{4.4}$$

where $x(t)$ denotes the value of the physical quantity 'position' at time t. This differential equation is second-order, and hence the value of x at time t_1 cannot alone fix the state since it does not determine uniquely the subsequent (or prior) motion.

However, this second-order equation can be rewritten as the pair of *first*-order equations for the physical quantities x and p (momentum)

$$\begin{aligned} m\frac{dx(t)}{dt} &= p(t) \\ \frac{dp(t)}{dt} &= F(x(t), \frac{p(t)}{m}) \end{aligned} \tag{4.5}$$

which, being first-order, are now deterministic. That is, the solution to this pair of equations is fixed uniquely once the values of x and p at some time t_1 have been specified. Thus the principle of causality would be

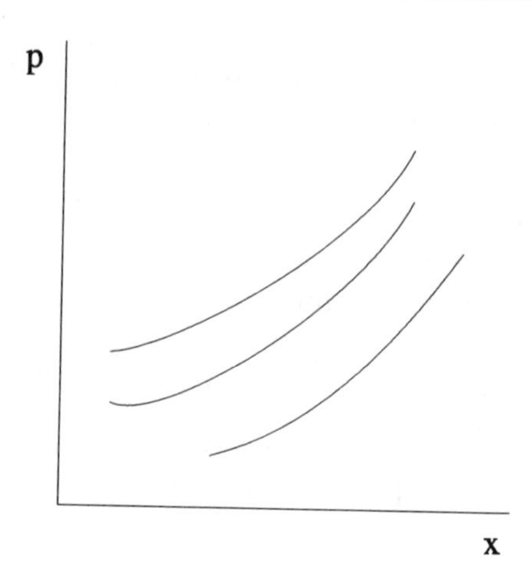

Figure 4.1: Trajectories in state space

satisfied if the state at any time t is defined to be the pair of real numbers $(x(t), p(t))$. Hence the space of states for this system may be regarded as the two-dimensional real vector space $\mathbb{R}^2$, with the possible histories of the system being represented by curves, as in Figure 4.1.

A physical quantity for the one-dimensional point particle is simply any function of x and p, and so the first criterion for the concept of a state is also satisfied. In particular, position and momentum are themselves represented by real-valued functions f_{pos} and f_{mom} on the state[9] space $\mathbb{R}^2$ defined by

$$f_{\text{pos}}(x,p) := x \tag{4.6}$$
$$f_{\text{mom}}(x,p) := p \tag{4.7}$$

for all pairs $(x,p) \in \mathbb{R}^2$.

Note that it has become standard to denote the functions f_{pos} and f_{mom} as x and p respectively, thus potentially confusing the *points* in the state space with these particular *functions* defined on it. Strictly speaking, these are not the same thing, and sometimes it is necessary to remember this. This confusion of points and functions is possible because f_{pos} and f_{mom} are coordinate functions on $\mathcal{S} = \mathbb{R}^2$ that project out the first and second components respectively of a pair (x,p). Note that f_{pos} and f_{mom} are state-fixing, *i.e.*, their joint values pick out a unique state.

[9] For historical reasons, the space $\mathbb{R}^2$ is often known as the *phase space* of the system.

The example of one-dimensional motion can be readily generalised. For example, a single particle moving in three dimensions has a state space $\mathbb{R}^6$, and the state space of N particles moving in three dimensions is $\mathbb{R}^{6N}$. More generally, the state space of any conventional classical system can be shown to be a space $\mathcal{S}$ with an even number of dimensions. In all cases, physical quantities are represented by real-valued functions on $\mathcal{S}$. Note that $\mathcal{S}$ does not have to be a vector space: for example, the state space of a particle constrained to move in a circle is a two-dimensional cylinder in which the distance along the axis represents the angular momentum of the particle as it moves round the circle.

In the discussion above, the mass m is not viewed as a function on the state space. Rather, its value, along with the functional form of the force F, constitutes part of the specification of the system itself. More precisely, the value of the mass is a 'structural' property—*i.e.*, one that is independent of the state of the system—whereas the values of x and p are 'contingent' since they are state dependent (and hence, in particular, can depend on time). However, both types of property are regarded as being *possessed* by the system. For any particular physical system there may be a number of structural quantities of this type. If desired, they also can be regarded as functions from $\mathcal{S}$ to $\mathbb{R}$ but with the added property that they are *constant* on all of $\mathcal{S}$. Quantities whose values are contingent then correspond to non-constant maps $f : \mathcal{S} \to \mathbb{R}$. These represent the 'dynamical degrees of freedom' of the system.

4.3 The Logical Structure of Classical Physics

The implications of these concepts of 'physical quantity', 'property', and 'state space' are captured nicely by the structure of the set of propositions associated with a classical system. By a 'proposition' I mean a statement about the properties of the system that is either true or false: for example, the statement that the energy lies between 1.2 and 1.7 ergs, or that a particle is a proton rather than a neutron, or that the position of the particle is 4.9 cms along some axis. Thus a typical proposition is of the form that the value of some physical quantity A lies in some interval Δ of the real numbers (after allowing for an appropriate choice of units), *i.e.*, the system *possesses* this property. Propositions of this type can be combined to give further propositions. For example, a pair of quantities A_1 and A_2 determine the compound proposition that the values of A_1 and A_2 lie in Δ_1 and Δ_2 respectively.

If $\mathcal{S}$ denotes the space of states of the system, and if the physical quantity A is represented by the function $f_A : \mathcal{S} \to \mathbb{R}$, then to the property that the value of A lies in Δ there is associated the subset[10]

$$\mathcal{S}_{A\in\Delta} := f_A^{-1}\{\Delta\} = \{s \in \mathcal{S} \mid f_A(s) \in \Delta\} \tag{4.8}$$

of all states in $\mathcal{S}$ for which the proposition that the system has this property is true. More generally, for *any* proposition T about properties of the system there is a corresponding subset $\mathcal{S}_T$ of states for which the proposition is true.

Any subset E of $\mathcal{S}$ determines a special *characteristic function* on $\mathcal{S}$ defined by

$$\chi_E(s) := \begin{cases} 1 & \text{if } s \in E, \\ 0 & \text{otherwise.} \end{cases} \tag{4.9}$$

If T is any proposition, we shall write χ_T for $\chi_{\mathcal{S}_T}$, *i.e.*, χ_T is the function on $\mathcal{S}$ which takes on the value 1 if T is true, and 0 otherwise. In particular, $\chi_{A\in\Delta}$ denotes the characteristic function associated with the proposition $A \in \Delta$ that the value of A lies in Δ. Thus $\chi_{A\in\Delta} : \mathcal{S} \to \mathbb{R}$ is defined by

$$\chi_{A\in\Delta}(s) := \begin{cases} 1 & \text{if } f_A(s) \in \Delta, \text{ } i.e., \text{ } s \in \mathcal{S}_{A\in\Delta} \\ 0 & \text{otherwise.} \end{cases} \tag{4.10}$$

Note that, viewed as a function from $\mathcal{S}$ to $\mathbb{R}$, a proposition can be thought of as a special type of physical quantity, *i.e.*, one whose numerical values are restricted to be 0 and 1.

It is clear that, in a certain sense, *any* subset E of $\mathcal{S}$ determines a particular proposition: namely that the state of the system lies in E. However, from a physical perspective this statement is rather abstract unless one can find a simple physical property, or collection of such, that pick out the particular subset E of states. In any event, we shall say that

[10]Strictly speaking, $\Delta \subset \mathbb{R}$ has to be a so-called *Borel* subset of the real numbers; similarly, the function $f_A : \mathcal{S} \to \mathbb{R}$ has to be a Borel function. It is also necessary to identify as equivalent any two subsets that differ by a set of Lebesgue measure zero. These restrictions rule out certain 'wild' subsets and functions; but it is not necessary to go into any further details here.

two propositions T_1 and T_2 are *physically equivalent* if they determine the same subset[11] of $\mathcal{S}$, *i.e.*, $\mathcal{S}_{T_1} = \mathcal{S}_{T_2}$ or, equivalently, $\chi_{T_1} = \chi_{T_2}$.

Note how this mathematical picture of a state space intertwines naturally with the classical approach to probability mentioned briefly in Section 2.1.1. For example, classical statistical physics works with a generalised volume function on the state space $\mathcal{S}$, so that we can associate the probability that the state[12] of the system lies in a subset E of $\mathcal{S}$ with the measure of E. In particular, for a probability density function ρ, the probability that the value of a quantity A lies in $\Delta \subset \mathbb{R}$ is (*cf.* Eq. (2.6))

$$\text{Prob}(A \in \Delta; \rho) = \int \cdots \int_{\mathcal{S}_{A\in\Delta}} \rho(x^1, x^2, \ldots, x^N)\, dx^1\, dx^2 \ldots dx^N \tag{4.11}$$

where N is the dimension of the space $\mathcal{S}$.

More generally, a *probability measure* on $\mathcal{S}$ is a map μ from the (Borel) subsets of $\mathcal{S}$ to the real numbers with the properties that

$$0 \le \mu(W) \le 1 \quad \text{for all such subsets } W, \tag{4.12}$$

$$\mu(\emptyset) = 0, \quad \mu(\mathcal{S}) = 1, \tag{4.13}$$

$$\mu(W_1 \bigcup W_2 \bigcup \cdots) = \mu(W_1) + \mu(W_2) + \cdots \tag{4.14}$$

where $\emptyset$ denotes the empty set, and where $W_1, W_2, \ldots$ is any finite, or countably infinite, collection of pairwise disjoint subsets of $\mathcal{S}$. The value of $\mu(W)$ is then interpreted as the probability that the state of the system is in the subset $W \subset \mathcal{S}$.

The association of physical equivalence classes of propositions with subsets of the state space $\mathcal{S}$ has very important implications for the logical structure of the set of propositions. For example, let T and U be two equivalence classes of propositions with corresponding subsets $\mathcal{S}_T$ and $\mathcal{S}_U$ respectively. Then:

1. The (equivalence class) of the proposition $T \wedge U$ ('T *and* U'; or the *conjunction* of T and U) is represented by the set-theoretic intersection $\mathcal{S}_T \bigcap \mathcal{S}_U$; *i.e.*, $\mathcal{S}_{T\wedge U} = \mathcal{S}_T \bigcap \mathcal{S}_U$. The associated characteristic functions satisfy

$$\chi_{T\wedge U} = \chi_T \chi_U \tag{4.15}$$

[11] Strictly speaking, if they determine the same subset of $\mathcal{S}$ modulo sets of Lebesgue measure zero.

[12] In classical statistical physics, an element of $\mathcal{S}$ is often called a *microstate*.

since $\chi_T(s)\chi_U(s) = 1$ if $\chi_T(s) = 1$ and $\chi_U(s) = 1$; otherwise $\chi_T(s)\chi_U(s) = 0$.

2. The proposition $T \vee U$ ('T *or* U'; or the *disjunction* of T and U) is represented by the union $\mathcal{S}_T \bigcup \mathcal{S}_U$; *i.e.*, $\mathcal{S}_{T\vee U} = \mathcal{S}_T \bigcup \mathcal{S}_U$. The associated characteristic functions satisfy

$$\chi_{T\vee U} = \chi_T + \chi_U - \chi_T\chi_U \tag{4.16}$$

since $\chi_T(s) + \chi_U(s) - \chi_T(s)\chi_U(s) = 0$ if $\chi_T(s) = 0$ and $\chi_U(s) = 0$; otherwise $\chi_T(s) + \chi_U(s) - \chi_T(s)\chi_U(s) = 1$.

3. The proposition $\neg T$ ('*not* T') is represented by the set-theoretic complement $\mathcal{S} - \mathcal{S}_T := \{s \in \mathcal{S} \mid s \notin \mathcal{S}_T\}$; *i.e.*, $\mathcal{S}_{\neg T} = \mathcal{S} - \mathcal{S}_T$. The associated characteristic functions satisfy

$$\chi_{\neg T} = 1 - \chi_T. \tag{4.17}$$

Another important concept is that of 'logical implication'. We say that an equivalence class of propositions T *implies* an equivalence class of propositions U if for all states for which T is true, U is also true. In set-theoretic terms this implies that $\mathcal{S}_T$ is a subset[13] of $\mathcal{S}_U$ or, equivalently, $\chi_T \leq \chi_U$, *i.e.*, for all $s \in \mathcal{S}$, $\chi_T(s) \leq \chi_U(s)$. We shall write this relation between equivalence classes of propositions as $T \preceq U$, and note that it satisfies the three conditions of a *partial ordering*:

1. for any proposition T, $T \preceq T$;

2. for any three propositions T, U and V, if $T \preceq U$ and $U \preceq V$ then $T \preceq V$;

3. if T and U are such that $T \preceq U$ and $U \preceq T$, then $T = U$.

Note that the proposition $\emptyset$ that is always false is identified with the empty set, and the proposition I that is always true is identified with the entire space $\mathcal{S}$. Thus, for any equivalence class of propositions T we have $\emptyset \preceq T \preceq I$.

An important property of the set of equivalence classes of propositions is that it possesses the structure of a *Boolean algebra* with respect to these

[13] As always, modulo sets of Lebesgue measure zero.

$\wedge$ and $\vee$ operations. In particular, the logical structure of the propositions is *distributive* in the sense that, for any three propositions T, U and V,

$$T \wedge (U \vee V) = (T \wedge U) \vee (T \wedge V) \tag{4.18}$$

$$T \vee (U \wedge V) = (T \vee U) \wedge (T \vee V) \tag{4.19}$$

as can be seen at once by drawing an appropriate Venn diagram [Exercise].

The crucial observation is that, by associating propositions in this way with subsets of the set $\mathcal{S}$, the logical structure of the *propositions* about the physical properties of the system is identified with the standard Boolean algebra structure on the *subsets* of the space of states $\mathcal{S}$. This is the precise way in which normal logical thinking (*i.e.*, Boolean logic) becomes implemented in the mathematical structure of classical physics.

4.4 Towards Quantum Theory

The general picture sketched above may seem to be in satisfactory accord with common sense, but in fact, under the impact of quantum ideas, every facet of it has come under review.

For example, in the conventional, rather pragmatic, approach to quantum theory, a sharp distinction is made between the system and the observer (or observing equipment). The primary emphasis is now placed on the act of *measurement*, with quantum theory being viewed as a scheme for predicting the probabilistic spread of the results obtained.

This rather instrumentalist view can be developed into a full-blown, anti-realist interpretation in which the idea of a physical system possessing values for all physical quantities is regarded as meaningless. From this perspective, it is *not* correct to say that a measurement yields a particular result because this is the value that the corresponding physical quantity happens to have at that time. In so far as quantum theory applies at all to single objects (and, to an anti-realist, this is debatable) a 'thing' is arguably better understood as a bundle of *latent*, or *potential*, properties that are only brought into being (in the sense of classical physics)[14] by the act of measurement. As we shall see, this failure of the classical notion

[14]These do not apply to 'structural' properties: they are always 'actual'.

of 'property' is reflected at a mathematical level in the *non*-distributive nature of the set of quantum propositions.

The concept of a state still plays a vital role in quantum theory, but, in the standard approach the first condition S1 can now be phrased only in the instrumentalist version of yielding precise predictions for the results of possible measurements. And, of course, since it is quantum theory we are discussing, it is the *probabilities* of getting certain results that are predicted, not the results themselves. The notion of the state as the bearer of the causal structure of the theory still stands and, in particular, a specification of the state at any one time is required to determine uniquely the state at any other time (provided no measurements are made between the two times); in this sense, condition S2 still holds. But what evolve deterministically are the predicted probabilities of measurement results, not the actual values of physical quantities.

These remarks need to be unpacked carefully if the full impact of quantum theory is to be appreciated. I shall attempt to do this at various points in what follows, but first we need a clear statement of the general formalism of quantum theory upon which later developments, be they technical or conceptual, can be based.

Chapter 5

THE GENERAL FORMALISM OF QUANTUM THEORY

5.1 The Rules of Quantum Theory

5.1.1 Prolegomenon

The rules of general quantum theory are generalisations of the four rules for using wave functions discussed in Chapter 1. They summarise the collective experience of the scientific community acquired since quantum theory was first formulated in the 1920s. Much of the remainder of this text is a logical unfolding of some of the mathematical, physical, and conceptual consequences that follow from them. However, it must be emphasised that these rules do not have the same status as, say, the axioms of a branch of pure mathematics (for example, the theory of vector spaces). In particular, at some time in the future an experimental result could be obtained that would require a substantial modification, or even total rejection, of the entire formalism.

The first problem we face is that it is not possible to state the structural rules for any theory of physics without assuming *something* about how they relate to the physical world. In a contentious subject such as quantum theory this raises the pedagogical question of how to present the material in a way that does not bias the conceptual issues too much in advance. I shall approach this problem by adopting initially a 'minimal' interpretative framework that most physicists are happy to accept, at least in a provisional way. Some of the problems that arise if one attempts to go beyond this framework will be discussed later. This minimal interpretation has the following key features:

- Quantum theory is viewed as a scheme for predicting the probabilistic distribution of the outcomes of *measurements* made on suitably-prepared copies of a system.

  To reflect this emphasis on measurement, the term 'physical quantity' gives way to 'observable quantity', or just 'observable'.

- The probabilities are interpreted in a statistical way as referring to the *relative frequencies* with which various results are obtained if the measurements are repeated a sufficiently large number of times.

- No claims are made about whether the invocation of the concept of 'measurement', or the emphasis placed on the relative-frequency interpretation of probability, are to be regarded as *fundamental* ingredients in the theory, or if they are merely a pragmatic reflection of what physicists actually do when carrying out their professional duties.

  In particular, nothing is said about whether the system 'possesses' values[1] for the physical quantities concerned before the measurements are made.

With its reluctance to address issues like the status of physical quantities, the minimal approach cannot be regarded as a full *interpretation* of the quantum formalism. On the other hand, it is undoubtedly the view adopted by many practicing physicists, and for this reason I shall refer to it as the *pragmatic approach* to quantum theory. In many respects it can

[1]Even in classical physics there are legitimate concerns about saying that a quantity has an exact value within a *continuum* of possible values, and this applies in quantum theory too. However, that is not the issue here. The central question is whether it is meaningful to say that an individual quantum system possesses values for those observables whose corresponding operators have only discrete eigenvalues. There is no problem in classical physics with a physical quantity whose values form a discrete set.

be thought of as the 'safe', fall-back position: moving towards a full-blown interpretation then resembles looking over a parapet towards the enemy lines whilst reserving the option to duck one's head at the first signs of fire!

In practice, physicists and philosophers who do jump the wall thereafter head off in a variety of directions—the underlying philosophies of which can be labelled broadly as either *anti-realist* or *realist*. An interpretation of the former type is characterised by some, or all, of the following positions:

- The notion of an individual physical system 'having', or 'possessing' values for all its physical quantities is *inappropriate* in the context of quantum theory. For the purposes of this book, this is a minimal requirement for any interpretation to be labelled as 'anti-realist'.

- The concept of 'measurement' is fundamental in the sense that the scope of quantum theory is *instrinsically* restricted to predicting the probabilistic spread of results of repeated measurements made on systems that have been prepared in precisely specified ways.[2] (Indeed, Bohr promulgated the idea that the very *concept* of a physical quantity can only be defined within the context of a particular measurement framework.) Thus interpretations of this type tend towards instrumentalist and operationalist philosophies of science.

- The relative-frequency interpretation of probability is a *fundamental* ingredient in quantum theory. In particular, the spread in the results of measurements on identically-prepared systems must *not* be interpreted as reflecting a 'lack of knowledge' of some objectively existing state of affairs.

- As a concomitant of the above, it may be claimed that a quantum state should not be associated with an *individual* system; rather, it refers only to a collection, or 'ensemble'[3] of copies of the system on

[2] A comprehensive discussion of this type of approach can be found in Peres (1993).

[3] The word 'ensemble' is often used by physicists to denote a collection of identically-prepared copies of a system on which measurements are to be made, and the probabilistic results interpreted in a relative-frequency sense. Historically, the concept of an ensemble was developed mainly in the context of classical statistical physics, where it is usually associated with an epistemic view of a system possessing unknown values for physical quantities. This is *not* the intended meaning here.

which repeated measurements are to be made.

Note that, notwithstanding the above, it *is* usually assumed that talk of an 'individual' system is meaningful (although this might be disputed by a complete instrumentalist, if there are any). In this sense, some properties—for example, spatial localisation—can be asserted of each copy of the individual system. Thus interpretations of this type tend to be anti-realist about most properties, but realist concerning certain others. In what follows, such interpretations will be qualified as 'relative-frequency', or 'instrumentalist', or 'anti-realist', according to the feature it is wished to emphasise.

At the opposite end of the spectrum lie those interpretations that aspire to return to the more realist view of classical physics. An interpretation of this type is characterised by some, or all, of the following positions:

- It *is* appropriate in quantum theory to say that an individual system possesses values for its physical quantities. In this context, 'appropriate' signifies that propositions of this type can be handled using standard propositional logic. For the purposes of this book, this is the minimal requirement for any interpretation to be labeled as 'realist'.

- Quantum theory is a framework for predicting the probabilistic distribution of these possessed values: it is *not* just a theory of the results of measurements. In particular, the concept of 'measurement' plays no *fundamental* role in the theory.

- Quantum-theoretical probabilities can be interpreted as a reflection of our lack of *knowledge* of what is actually the case: this is the so-called *epistemic* interpretation of probability.

- It *is* appropriate to assign quantum states to an individual system. Some physicists interpret this as saying that the state refers to the *system* itself. Another view is that a state refers to our *knowledge* of the system.

A realist position of this type is often accompanied by a claim that the current technical formalism of quantum theory is *incomplete*. In particular, 'hidden variables' exist whose values are needed to complete the physical description. Two contrasting views here are that (i) the values of these hidden variables are determinable in principle, and hence we should be able to get back to a strict deterministic view of nature; and (ii) the hidden variables are subject to fundamental statistical fluctuations which underlie, and validate, the probabilistic nature of the predictions of standard quantum theory.

The many positions that exist between the extreme realist and anti-realist stances can be classified from a variety of perspectives. For example, a classification scheme could be based on the assumed referent of a quantum state, some of the options for which include:

- an individual quantum system;
- our *knowledge* of the properties of such a system;
- the result of any *measurement* that could be made on such a system;
- a *collection* (real or hypothetical) of identically-prepared copies of the system on which repeated measurements are to be made;
- the results of *repeated measurements* that could be made on such a collection.

Another possibility is to focus on the precise meaning of 'probability' when used in the context of quantum theory.

A third option, and the one favoured here, is to probe the extent to which it is, or is not, possible to regard individual systems as *possessing* properties in typical quantum situations. The importance of this question was emphasised by Margenau and his colleagues in a series of papers (Margenau 1949, McKnight 1952, Margenau 1963*a*, Margenau 1963*b*, Margenau 1963*c*, Park & Margenau 1968, Park 1968*a*, Park 1968*b*). It is also one of the central themes in Redhead (1989).

For example, one extension of the minimal approach is to say that an individual system *can* possess values, but these values are intrinsically 'unsharp' in some way (Busch, Lahti & Mittelstaedt 1991). Another idea is that an individual system does not possess *values* of physical quantities but rather—to adopt Sir Karl Popper's term—*propensities* to have values (Popper 1956). In one version of this idea, the values of quantities are 'latent' and 'come into being' upon measurement. In a more realist view, a propensity to have a certain value is itself regarded as a fundamental property of a physical system. The idea of latent values was discussed extensively by Margenau and colleagues.

5.1.2 Statement of the Rules

These few remarks raise many difficult issues, and we shall return later to some of the questions that arise when the conceptual framework of quantum theory is probed more closely. However, most of the remainder of this chapter is concerned with the principle technical ingredients of quantum theory as viewed from within the minimal, pragmatic approach.

The four rules that follow deal with the general mathematical framework[4] within which it has been found possible so far to describe *all* quantum-mechanical systems.

Rule 1. The predictions of results of measurements made on an otherwise isolated system are probabilistic in nature. In situations where the maximum amount of information is available, this probabilistic information is represented mathematically by a vector in a complex Hilbert space $\mathcal{H}$ that forms the *state space* of the quantum theory. In so far as it gives the most precise predictions that are possible, this vector is to be thought of as the mathematical representative of the physical notion of 'state' of the system.

Rule 2. The observables of the system are represented mathematically by self-adjoint operators that act on the Hilbert space $\mathcal{H}$.

Rule 3. If an observable quantity A and a state are represented respectively by the self-adjoint operator $\widehat{A}$ and the normalised vector[5]

[4]There are other mathematical frameworks within which quantum theory can be developed. A well-known example, often used in rigorous quantum field theory, takes as its central object a certain algebraic structure (a C^*-algebra) on the set of physical observables. States are then defined in relation to this algebra. This contrasts with the Hilbert-space approach in which the primary object is the vector space of states, with observables being defined in relation to this space.

[5]A vector $\vec{\psi}$ is *normalised* if $\langle\vec{\psi},\vec{\psi}\rangle = 1$.

$\vec{\psi} \in \mathcal{H}$, then the expected result[6] $\langle A \rangle_\psi$ of measuring A is

$$\langle A \rangle_\psi = \langle \vec{\psi}, \hat{A}\vec{\psi} \rangle \tag{5.1}$$

or, in Dirac notation,

$$\langle A \rangle_\psi = \langle \psi | \, \hat{A} \, | \psi \rangle. \tag{5.1'}$$

Rule 4. In the absence of any external influence (*i.e.*, in a *closed* system), the state vector $\vec{\psi}_t$ changes smoothly in time t according to the time-dependent Schrödinger equation

$$i\hbar \frac{d\vec{\psi}_t}{dt} = \widehat{H}\vec{\psi}_t \tag{5.2}$$

where $\widehat{H}$ is a special operator known as the *Hamiltonian*.

5.1.3 Some Comments on Rules 1, 2 and 3

1. Rules 1 and 2 are *correspondence* rules that specify the way in which quantities pertaining to the physical world are to be represented mathematically. As emphasised already, they are phrased in terms of the results of making *measurements* on the system rather than in terms of possessed *properties*. This 'black-box' perspective is part of the minimal, pragmatic approach to quantum theory and, although there are frequent references in what follows to a 'state of a system', this must be understood as relating only to what will happen on the average in a long series of repeated measurements.

As mentioned above, some physicists like to reinforce this by insisting that the state vector refers only to the large *collection* of suitably-prepared systems on which repeated measurements are to be made. I will generally adopt this terminology whenever I wish to emphasise the role of repeated measurements. However, to some extent, this is a matter of convention.

[6] In order to comply with the conventions of standard probability theory, the word 'average' is best reserved for the average of an actual series of measurements. When referring to the 'average' predicted by the mathematical formalism, it is more appropriate to use the phrase 'expected result' or, in a more realist form, 'expected value'.

The concept of a 'state' of a system is rather abstract and, as such, does not have the same intuitive meaning in normal discourse as does, for example, a statement that an object has a certain position. From this perspective, it is arguably as meaningful to say that a state refers to a single system—but only in so far it leads to probabilistic predictions of the results of repeated measurements—as it is to associate the state with the collection of systems on which the measurements are to be made.

Indeed, one could argue that the first statement is actually preferable since it can be read as relating to the results of an *infinite* sequence of hypothetical measurements, whereas any *actual* collection of copies of a system on which measurements are made, will always be *finite* in number, and hence the relative frequencies of the outcomes will only approximate the theoretical probabilities. It is also worth noting that actual acts of state *preparation* (see below) are almost always performed on individual systems. Thus, an instrumentalist-minded physicist might assert that to say that a system is in a certain state means precisely that it has been subjected to a certain preparation procedure. In this sense, it is quite natural to associate a state with an individual system.

The really important point is that, unlike the analogous situation in classical physics, to specify a quantum state is *not* to assign definite values to all physical quantities. The danger in referring a state to an individual system is of being lured thereby into thinking that physical quantities necessarily have values for each such system: referring the state to a collection of systems tends to avoid this trap.

We shall return later to the feasibility of interpreting quantum theory in a more realist way in which meaning *is* given to the notion of the properties of an individual system. Note that, in any such interpretation, one would probably want to replace Rule 3 with something like:

> "If a physical quantity A and a state of the system are represented respectively by the self-adjoint operator $\widehat{A}$ and the normalised vector $\vec{\psi} \in \mathcal{H}$, then the expected value $\langle A\rangle_\psi$ of A is $\langle A\rangle_\psi = \langle\vec{\psi}, \widehat{A}\vec{\psi}\rangle$."

In particular, "expected result $\langle A\rangle_\psi$ of measuring A" is replaced with "expected value $\langle A\rangle_\psi$ of A".

2. The notion of repeated measurements plays an important role in the pragmatic approach to quantum theory, but it is a little vague. In practice, repeated measurements may be made on many copies of the same system, all of which are measured at essentially the same time (for example, when shining a laser into a box of atoms to study the atomic levels); or the copies may be measured sequentially (for example, if an atomic beam is passed through a Stern–Gerlach apparatus); or the experiment may be repeated many times using the same system suitably prepared on each occasion. But in all cases two important questions arise.

The first is knowing how many times an experiment should be repeated in order that the measured relative frequencies of different outcomes be an acceptable approximation to the theoretically predicted probabilities. This issue arises whenever measurements are made in physics, and is often assumed not to generate any considerations peculiar to quantum theory. However, this is not obvious, since standard statistical analysis is predicated on classical probability theory, which is not appropriate in quantum theory. There have been some interesting attempts to tackle this issue by seeing what happens if a large number of copies of a system are treated as a *single* composite system within the framework of quantum theory. The aim is to define 'relative-frequency' operators on this composite system, whose eigenvalues are the relative frequencies of measurement results on the constituent systems. A well-known example is Hartle (1968); for a recent brief summary see Busch et al. (1991).

The second issue that we must address is how to guarantee that the appropriate copy of the system is in the 'right state' at the time of making a repeated measurement: a query that is particularly problematic if posed within any interpretation that is reluctant to associate quantum states with individual systems. The standard resolution of this problem of reproducible acts of state preparation will be discussed in Chapter 8.

3. Not every vector in $\mathcal{H}$ necessarily represents a state that can actually be realised. In particular, this is so for vectors that are superpositions of eigenstates of certain operators. One important example is electric charge; another is fermion number. The non-existence of such states in nature is reflected in the mathematical formalism by the existence of so-called *superselection rules*. This is an area in which the alternative, C^*-algebra, approach to quantum theory is especially useful.

4. In the early axiomatisations of quantum theory it was assumed that every self-adjoint operator represents some physical observable. However, for most operators it is very difficult to imagine pieces of equipment that could measure the corresponding observable and, in practice, most applications of quantum theory involve only a small number of physical quantities such as position, linear momentum, angular momentum, energy, *etc.* In addition, the existence of superselection rules suggests strongly that operators that project onto inadmissible vectors cannot represent physical observables.

5. Not all physical observables are associated with non-trivial self-adjoint operators. For example, as mentioned in Section 4.2, in Newtonian physics the mass m is a structural quantity. This is reflected in the corresponding quantum theory, where m appears as a parameter in the Hamiltonian, not as a non-trivial operator. However, this feature is theory-dependent since, in more sophisticated quantum accounts, the mass m can become an operator whose eigenvalues yield, for example, the masses of the elementary particles.

6. The states discussed above correspond to the situation in which we have the best possible 'information' about the quantum system. However, in practice this may be far from the case, and often the best that can be said is that the state is drawn from some collection $\vec{\psi}_1, \vec{\psi}_2, \ldots, \vec{\psi}_D$ (an infinite value for D is permitted), and that the probability of the state being $\vec{\psi}_d$, $d = 1, 2, \ldots, D$, is some real number w_d with $0 < w_d \leq 1$ and $\sum_{d=1}^{D} w_d = 1$. This uncertainty of which vector state is actually realised is to be interpreted in a purely classical way. It could arise, for example, from the vagaries of the equipment used to prepare the system; or perhaps from unavoidable thermal fluctuations induced by a non-zero temperature environment that cannot be isolated from the quantum system.

The collection $(\vec{\psi}_1, \vec{\psi}_2, \ldots, \vec{\psi}_D;\ w_1, w_2, \ldots, w_D)$ is known as a *mixed state* of the system, and we shall discuss in the next chapter how Rule 3 must be modified in order to combine the purely classical probabilities $w_1, w_2, \ldots, w_D$ with the intrinsic quantum probabilities attached to the vector states $\vec{\psi}_1, \vec{\psi}_2, \ldots, \vec{\psi}_D$.

7. Equation Eq. (5.1) implies that the physical predictions of the theory are unchanged if the state vector $|\psi\rangle$ is multiplied by an arbitrary complex number μ such that $|\mu| = 1$.

8. Sometimes the quantum rules are stated using state vectors that are not necessarily normalised. In this case Rule 3 becomes

$$\langle A \rangle_\psi = \frac{\langle \psi | \widehat{A} | \psi \rangle}{\langle \psi | \psi \rangle}. \tag{5.3}$$

5.2 Quantisation of a Given Classical System

5.2.1 Preservation of Classical Structure

The rules above are very general and need to be supplemented with additional assumptions that depend on the type of system under discussion. This applies in particular to the choice of an *actual* Hilbert space for a given physical system, and the identification of specific operators with specific observable quantities.

One very important situation in practice is when the task is to construct the 'quantum analogue' of a *given* classical system with a specific state space $\mathcal{S}$. A large number of genuine quantum systems are obtained in this way, ranging from 'quantising' the harmonic oscillator or the hydrogen atom, to constructing the quantum theory of unified electro-weak interactions.

In the classical theory, each physical quantity is associated with a (Borel) function $f : \mathcal{S} \to \mathbb{R}$; 'quantisation' can then be viewed as a map $f \mapsto \widehat{f}$ that associates to each such function f a self-adjoint operator $\widehat{f}$ on the quantum state space $\mathcal{H}$. For example, consider the familiar symbols $\widehat{x}$ and $\widehat{p}$ that arise in the wave mechanics of a point particle moving in one dimension. The classical state space is the set $\mathbb{R}^2$ of all pairs (x, p), and, as explained in Section 4.2.2, position and momentum are represented by the functions f_{pos} and f_{mom} defined in Eq. (4.6–4.7). Thus, from this perspective, the symbols $\widehat{x}$ and $\widehat{p}$ are a shorthand for $\widehat{f}_{\text{pos}}$ and $\widehat{f}_{\text{mom}}$ respectively.

The importance of this way of looking at things is that the set $C(\mathcal{S}, \mathbb{R})$ of all such classical functions $f : \mathcal{S} \to \mathbb{R}$ is highly structured, and the possibility arises that some of this will be preserved by the quantisation map $f \mapsto \widehat{f}$, thus giving valuable insight into the whole quantisation procedure.

The space $C(\mathcal{S}, \mathbb{R})$ carries two[7] natural mathematical structures. The first is given by the usual vector operations Eq. (2.25–2.26) on any space of vector-valued functions. Namely, if $f_1, f_2 \in C(\mathcal{S}, \mathbb{R})$ and $r_1, r_2 \in \mathbb{R}$,

$$(r_1 f_1 + r_2 f_2)(s) := r_1 f_1(s) + r_2 f_2(s) \tag{5.4}$$

for all $s \in \mathcal{S}$. By this means, $C(\mathcal{S}, \mathbb{R})$ becomes a real vector space.

The second natural structure comes from defining the *product* $f_1 f_2$ of two functions $f_1, f_2 \in C(\mathcal{S}, \mathbb{R})$ by

$$(f_1 f_2)(s) := f_1(s) f_2(s) \tag{5.5}$$

for all $s \in \mathcal{S}$.

For our purposes, the crucial question is whether any of this algebraic structure is preserved by quantisation. For example, if classical observables f_1 and f_2 are represented in the quantum theory by operators $\widehat{f_1}$ and $\widehat{f_2}$ respectively, is the linear combination $r_1 f_1 + r_2 f_2$ represented by $r_1 \widehat{f_1} + r_2 \widehat{f_2}$? In other words, is the quantisation operation $f \mapsto \widehat{f}$ a *linear* map from $C(\mathcal{S}, \mathbb{R})$ to the real vector space of self-adjoint operators on the quantum state space $\mathcal{H}$?

In this context it is instructive to consider how, for example, the simple harmonic oscillator is handled in elementary wave mechanics; in particular, the way in which the operator that represents the energy is obtained. The classical energy function on the state space $\mathcal{S} = \mathbb{R}^2$ is

$$H(x, p) := \frac{p^2}{2m} + \frac{m\omega^2}{2} x^2 \tag{5.6}$$

[7] A third, very important, structure is that $\mathcal{S}$ is usually a *symplectic* manifold. This gives rise to the 'Poisson bracket' that associates with each pair of functions $f, g : \mathcal{S} \to \mathbb{R}$, a third function denoted $\{f, g\}$. The Poisson bracket plays a fundamental role in the generalised Hamiltonian and Lagrangian approaches to classical mechanics, and Dirac made it central in his approach to quantisation. In particular, he postulated that the quantum operator associated with the classical function $\{f, g\}$ is proportional to the commutator $[\widehat{f}, \widehat{g}]$ of the operators representing f and g. This idea plays a major part in many modern approaches to the quantisation of classical systems, especially those with a classical state space $\mathcal{S}$ that is topologically complex.

and the corresponding operator in quantum theory is assumed to be

$$\widehat{H} = \frac{\widehat{p}^2}{2m} + \frac{m\omega^2}{2}\widehat{x}^2 \tag{5.7}$$

which acts on wave functions $\psi(x)$ via the definitions

$$(\widehat{x}\psi)(x) := x\psi(x) \tag{5.8}$$

$$(\widehat{p}\psi)(x) := -i\hbar\frac{d\psi}{dx}(x) \tag{5.9}$$

so that

$$(\widehat{H}\psi)(x) := \frac{-\hbar^2}{2m}\frac{d^2\psi(x)}{dx^2} + \frac{m\omega^2}{2}x^2\psi(x). \tag{5.10}$$

We shall return later to the origin of Eq. (5.8–5.9), but for the moment let us concentrate on how Eq. (5.7) is obtained from Eq. (5.6). We begin by rewriting Eq. (5.6) as a specific relation between functions on the classical state space:

$$H = \frac{1}{2m}(f_{\text{mom}})^2 + \frac{m\omega^2}{2}(f_{\text{pos}})^2 \tag{5.11}$$

and then look carefully at how the map $H \mapsto \widehat{H}$ is constructed. A little thought shows that the following (often implicit) assumptions are made:

1. The operator that represents the classical function $(f_{\text{mom}})^2$ is $\widehat{p}^2$, *i.e.*, the square of the operator that represents f_{mom}; and analogously for f_{pos}.
2. The operator that represents $\frac{1}{2m}(f_{\text{mom}})^2$ is $\frac{1}{2m}$ times the operator that represents $(f_{\text{mom}})^2$; and analogously for $\frac{m\omega^2}{2}(f_{\text{pos}})^2$.
3. The operator that represents the sum of $\frac{1}{2m}(f_{\text{mom}})^2$ and $\frac{m\omega^2}{2}(f_{\text{pos}})^2$ is the sum of the operators that represent $\frac{1}{2m}(f_{\text{mom}})^2$ and $\frac{m\omega^2}{2}(f_{\text{pos}})^2$.

Such operations are performed frequently when using wave mechanics, and are encapsulated in the following two assumptions about the quantisation procedure:

Q1 The quantisation procedure is *linearity preserving*, *i.e.*, the quantisation map $f \mapsto \widehat{f}$ is linear, so that

$$r_1 f_1 + r_2 f_2 \mapsto r_1\widehat{f_1} + r_2\widehat{f_2}. \tag{5.12}$$

Q2 The quantisation procedure is *function preserving* in the following sense. If $\mathcal{F} : \mathbb{R} \to \mathbb{R}$ is any real function, the real-valued function $\mathcal{F}(f)$ on $\mathcal{S}$ is defined as

$$\mathcal{F}(f)(s) := \mathcal{F}(f(s)) \qquad \text{for all } s \in \mathcal{S}. \tag{5.13}$$

Then, if the operator that represents f is $\widehat{f}$, the operator that represents $\mathcal{F}(f)$ is $\mathcal{F}(\widehat{f})$, *i.e.*,

$$\widehat{\mathcal{F}(f)} = \mathcal{F}(\widehat{f}). \tag{5.14}$$

A common assumption in quantum theory in general (*i.e.*, not just in wave mechanics) is that the quantisation of any given classical system obeys these two rules. Of course, to be meaningful, this requires a definition of the symbol $\mathcal{F}(\widehat{f})$ for a general function $\mathcal{F}$. This will be given in Section 5.2.2.

Comments

1. Conventional wave mechanics is obtained by adding the additional requirement that the operators $\widehat{x}$ and $\widehat{p}$, which represent position and momentum respectively, should satisfy the *canonical commutation relation*

$$[\,\widehat{x}, \widehat{p}\,] = i\hbar. \tag{5.15}$$

As we shall show in Section 7.2, this is deeply connected with the homogeneous nature of physical space.

It is also required that every operator on the Hilbert space can be written as a function of the operators $\widehat{x}$ and $\widehat{p}$. This is intended to reflect the fact that, in the corresponding classical theory, every observable is a function of x and p (*i.e.*, a function of the functions f_{pos} and f_{mom}). A remarkable theorem due to Stone and von Neumann shows that (with certain technical caveats) there is essentially only one Hilbert space on which self-adjoint operators can be found satisfying Eq. (5.15); any other is guaranteed to give the same physical results. This special space is the Hilbert space $L^2(\mathbb{R})$ of square-integrable functions with the inner product Eq. (2.43) (*i.e.*, the usual overlap function) and the familiar operators in Eq. (5.8–5.9) for $\widehat{x}$ and $\widehat{p}$ (Reed & Simon 1972).

2. An additional assumption sometimes made in elementary wave mechanics is that if $f(x, p)$ is any classical observable, then the corresponding

quantum operator is $f(\widehat{x}, \widehat{p})$. However, note that the latter is ill-defined since, for example, although xp and px are the same classical observable, the operators $\widehat{x}\widehat{p}$ and $\widehat{p}\widehat{x}$ are different (because $\widehat{x}$ and $\widehat{p}$ do not commute). In addition, neither of them is self-adjoint. Thus this, so-called, *substitution rule* needs to be used very cautiously.

Certain substitution laws can be derived from the linearity-preserving and function-preserving assumptions. For example, according to these rules, the quantisation[8] of $(x+p)^2$ is $(\widehat{x}+\widehat{p})^2 = \widehat{x}^2 + \widehat{p}^2 + \widehat{x}\widehat{p} + \widehat{p}\widehat{x}$. On the other hand, $(x+p)^2 = x^2 + p^2 + 2xp$, and the quantisation of x^2 and p^2 is $\widehat{x}^2$ and $\widehat{p}^2$ respectively. Hence, by writing $2xp \equiv (x+p)^2 - x^2 - p^2$, it can be argued that $2xp \mapsto (\widehat{x}^2 + \widehat{p}^2 + \widehat{x}\widehat{p} + \widehat{p}\widehat{x}) - \widehat{x}^2 - \widehat{p}^2$, so that

$$xp \mapsto \frac{1}{2}(\widehat{x}\widehat{p} + \widehat{p}\widehat{x}) \tag{5.16}$$

i.e., $\widehat{xp} = \frac{1}{2}(\widehat{x}\widehat{p} + \widehat{p}\widehat{x})$.

The argument used above is easily extended to show that linearity-preserving and function-preserving assumptions imply that, for *any* pair of functions $f, g : \mathcal{S} \to \mathbb{R}$,

$$fg \mapsto \frac{1}{2}(\widehat{f}\widehat{g} + \widehat{g}\widehat{f}). \tag{5.17}$$

However, this particular quantisation rule leads to a series of contradictions. For example, one could argue from Eq. (5.16–5.17) that the quantisation of $x(xp)$ is

$$\begin{aligned} x(xp) &\mapsto \frac{1}{2}(\widehat{x}\,\widehat{xp} + \widehat{xp}\,\widehat{x}) = \frac{1}{4}(\widehat{x}(\widehat{x}\widehat{p} + \widehat{p}\widehat{x}) + (\widehat{x}\widehat{p} + \widehat{p}\widehat{x})\widehat{x}) \\ &= \frac{1}{4}(\widehat{x}^2\,\widehat{p} + \widehat{p}\,\widehat{x}^2 + 2\widehat{x}\,\widehat{p}\,\widehat{x}). \end{aligned} \tag{5.18}$$

On the other hand, using the same assumptions, the quantisation of $(x^2)p$ (which, classically, is identical to $x(xp)$) should be

$$(x^2)p \mapsto \frac{1}{2}(\widehat{x^2}\,\widehat{p} + \widehat{p}\,\widehat{x^2}) = \frac{1}{2}(\widehat{x}^2\,\widehat{p} + \widehat{p}\,\widehat{x}^2), \tag{5.19}$$

[8]Strictly speaking, we should talk of the quantisation of the function $(f_{\text{pos}} + f_{\text{mom}})^2$. However, we will revert to the physicist's convention of writing this as $(x+p)^2$, *i.e.*, we specify a function in terms of its value at a general point (x, p) in the classical state space.

which is not the same as Eq. (5.18).

Such contradictions show that the linearity-preserving and function-preserving quantisation assumptions cannot be invoked as genuine rules. They can at best be regarded as a heuristic way of constructing certain quantum operators starting with a classical system, but they cannot be given universal status. Alternatively, if they *are* taken as fundamental, then one of the basic quantum rules must be changed. The usual assumption is that it is *not*, in fact, possible to associate every function $f : \mathcal{S} \to \mathbb{R}$ with a quantum operator. Instead, when faced with the task of quantising a given classical system, one starts by associating self-adjoint operators with only a small subset of classical observables (for example, in wave mechanics, this could be just linear sums of x and p) and then tries cautiously to extend this procedure to include further physical quantities of interest.

3. The need for such a tentative approach is rather unsatisfactory, and suggests that the whole idea of 'quantising' a given classical system is suspect even though, in practice, the procedure does generate many quantum systems of considerable importance. Arguably, one is allowing Bohr's strictures on the need to invoke 'classical' measuring devices in the *interpretation* of quantum theory to limit the actual construction of such theories. A more logical process would be to start from a quantum system that is given in some intrinsic way, and then to ask about its classical limit.

Unfortunately, there is no clear understanding of what it means to specify a quantum system in an 'intrinsic way', and an important question therefore is whether there are analogues in this case of the linearity-preserving and function-preserving assumptions that arose in the context of quantising a given classical system.

In the absence of any classical system that is being quantised, it is tempting to think of 'quantisation' as a map $A \mapsto \widehat{A}$ that is defined directly on physical quantities, rather than on functions on a classical state space. From this perspective, the letter A is merely shorthand for the name of a physical quantity of interest—'energy', 'position', 'angular momentum', *etc.*—in which case the association $A \mapsto \widehat{A}$ is a map from the, rather unstructured, class of all such quantities to the set of self-adjoint operators on a Hilbert space $\mathcal{H}$. If there is no underlying classical structure, this quantisation map cannot be seen as 'factoring' through an intermediate function $f_A : \mathcal{S} \to \mathbb{R}$ and a quantisation $f_A \mapsto \widehat{A} := \widehat{f_A}$, *i.e.*, we have $A \mapsto \widehat{A}$ but not $A \mapsto f_A \mapsto \widehat{f_A}$. The central question is what sense—if any—can now be made of quantisation conditions like $r_1 A_1 + r_2 A_2 \mapsto r_1 \widehat{A_1} + r_2 \widehat{A_2}$, or $\widehat{\mathcal{F}(A)} = \mathcal{F}(\widehat{A})$.

The main issue here is the physical meaning of mathematical expressions like $r_1 A_1 + r_2 A_2$ or $\mathcal{F}(A)$. For example, in classical physics, $\mathcal{F}(A)$ denotes that physical quantity whose *value* in any state is the appropriate function $\mathcal{F}$ of the value of A. Thus, in terms of the function $f_A : \mathcal{S} \to \mathbb{R}$ that represents the physical quantity A, the quantity $\mathcal{F}(A)$ is defined by specifying the associated function $f_{\mathcal{F}(A)} : \mathcal{S} \to \mathbb{R}$ to be

$$f_{\mathcal{F}(A)}(s) := \mathcal{F}(f_A(s)) \tag{5.20}$$

for all $s \in \mathcal{S}$. However, such a definition is not appropriate within the pragmatic approach to quantum theory since it is not assumed that an observable possesses a value before it is measured. Indeed, as we shall see in the discussion of the Kochen–Specker theorem (Section 9.1), the interpretation of the symbol $\mathcal{F}(A)$ is an important issue even in approaches that do try to assign possessed properties to quantum systems.

One approach to giving meaning to a function $\mathcal{F}$ of an observable A is to adopt an *operational* definition of the observable $\mathcal{F}(A)$ by specifying how it is to be measured: *i.e.*, measure A and then apply the function $\mathcal{F}$ to the result of the measurement. It then becomes meaningful to impose the function-preserving requirement Eq. (5.14). Similarly, there is no problem in finding an operational definition of a product rA of an observable A with a real number r: it is simply that quantity which is measured by measuring A, and then multiplying the result by r.

The situation is somewhat different for a sum $A + B$ of observables A and B. The operationalist strategy would require us to give meaning to $A + B$ by measuring A and B simultaneously and then to take the sum of the results. However, a problem arises if the operators $\widehat{A}$ and $\widehat{B}$ that represent A and B do not commute. Many people take a non-vanishing commutator to indicate that A and B cannot be measured simultaneously, in which case no such operational meaning of $A + B$ can be given. Whether or not this is really the case depends on precisely what constitutes a 'measurement'. We shall return to this issue in Section 6.3, but for the moment it suffices to note that Park and Margenau have suggested several ways in which noncommuting quantities *could* be said to be measured simultaneously (Park & Margenau 1968). However, they then go on to demonstrate the type of inconsistency mentioned above in the context of Eq. (5.18–5.19).

Park and Margenau gave a nice example of simultaneous measurement, considering the x and y components of electron spin, where the corresponding quantum operators are $\widehat{S}_x = \frac{\hbar}{2}\sigma_x$ and $\widehat{S}_y = \frac{\hbar}{2}\sigma_y$ and where σ_x and σ_y are the Pauli spin matrices defined in Eq. (1.23). The result[9] of measuring either S_x or S_y is $\pm\frac{\hbar}{2}$, and hence the sum of the results of a simultaneous measurement, could any such be made, must be either $\hbar$, 0, or $-\hbar$. On the other hand, the operator $\frac{\hbar}{2}\sigma_x + \frac{\hbar}{2}\sigma_y$ is $\frac{\hbar}{2}\begin{pmatrix} 1 & 1 \\ 1 & -1 \end{pmatrix}$, and this has eigenvalues $\hbar/\sqrt{2}$ and $-\hbar/\sqrt{2}$, which are not equal to $\hbar$, 0 or $-\hbar$. Thus there is a contradiction if one thinks that the operationalist meaning of $A+B$ involves measuring A and B simultaneously and taking the sum of the results.

The existence of such a contradiction is one of the features that distinguishes quantum physics from classical physics. For example, the value of the classical Hamiltonian Eq. (5.11) in a state (x,p) can be obtained by taking the sum of the values of $\frac{1}{2m}(f_{\text{mom}})^2$ and $\frac{m\omega^2}{2}(f_{\text{pos}})^2$ in that state. However, the eigenvalues of the quantum operator $\widehat{H}$ in Eq. (5.7) are not related in any simple way to those of the operators $\frac{\widehat{p}^2}{2m}$ and $\frac{m\omega^2}{2}\widehat{x}^2$. Thus, even if position and momentum could be measured together, the possible results of measuring energy directly in quantum theory would not be equal to the sums of the results of measuring $\frac{p^2}{2m}$ and $\frac{m\omega^2}{2}x^2$.

Margenau and Park concluded from such arguments that, in general (not just in the case where there is an underlying classical system that is being quantised), not every observable quantity corresponds to a quantum operator. Thus they would amend Rule 2 to read that *some* observables are associated with self-adjoint operators. In any event, it is clear that great care needs to be taken when talking about algebraic functions of observables. In practice, the most important condition is the function-preserving requirement $\widehat{\mathcal{F}(A)} = \mathcal{F}(\widehat{A})$ which *is* normally assumed to hold in any quantisation scheme. We shall return to this issue in Section 9.1

[9] I am assuming here that the result of measuring an observable A must be one of the eigenvalues of the associated self-adjoint operator $\widehat{A}$. This will be proved in Section 5.3.

when we discuss the Kochen-Specker theorem.

5.2.2 The Definition of $\mathcal{F}(\widehat{A})$

The next problem is to interpret the expression $\mathcal{F}(\widehat{A})$ that has been used freely in the above; *i.e.*, what is meant by a function of an operator? Polynomial functions of any operator $\widehat{A}$ are easily defined; for example, for $b\widehat{A} + c\widehat{A}^2 + d\widehat{A}^3$ we have $(b\widehat{A} + c\widehat{A}^2 + d\widehat{A}^3)\vec{\psi} := b(\widehat{A}\vec{\psi}) + c\widehat{A}(\widehat{A}(\vec{\psi})) + d\widehat{A}(\widehat{A}(\widehat{A}(\vec{\psi})))$. But what is meant by an expression like $\log \widehat{A}$?

In the case when $\widehat{A}$ is self-adjoint this question has a well-known, and very important, answer. Suppose first (using Dirac notation) that $|a\rangle$ is an eigenstate of $\widehat{A}$ with eigenvalue a, *i.e.*, $\widehat{A}\,|a\rangle = a\,|a\rangle$. Then

$$\widehat{A}^2\,|a\rangle = a^2\,|a\rangle, \qquad \widehat{A}^3\,|a\rangle = a^3\,|a\rangle \tag{5.21}$$

and, in general, if Q is any polynomial function we have

$$Q(\widehat{A})\,|a\rangle = Q(a)\,|a\rangle. \tag{5.22}$$

Motivated by Eq. (5.22) we define for an *arbitrary* function $\mathcal{F} : \mathbb{R} \to \mathbb{R}$,

$$\mathcal{F}(\widehat{A})\,|a\rangle := \mathcal{F}(a)\,|a\rangle. \tag{5.23}$$

For example, $\log \widehat{A}\,|a\rangle := \log a\,|a\rangle$, which makes sense if the eigenvalue a is positive. In this way, meaning is given to $\mathcal{F}(\widehat{A})$ acting on any eigenvector of $\widehat{A}$, provided the function $\mathcal{F}$ is well-defined on the corresponding eigenvalue.

To extend this to include *all* vectors we exploit the basic expansion theorem Eq. (3.51)

$$|\psi\rangle = \sum_{m=1}^{M} \sum_{j=1}^{d(m)} \langle a_m, j|\psi\rangle\,|a_m, j\rangle \tag{5.24}$$

where $a_1, a_2, \ldots, a_M$ (M could be ∞) label the distinct eigenvalues of $\widehat{A}$ (whose spectrum is assumed to be discrete) and $j = 1, 2, \ldots, d(m)$ labels the degenerate eigenvectors with the common eigenvalue a_m. Then equation Eq. (5.23) suggests the definition

$$\mathcal{F}(\widehat{A})\,|\psi\rangle := \sum_{m=1}^{M} \sum_{j=1}^{d(m)} \mathcal{F}(a_m)\langle a_m, j|\psi\rangle\,|a_m, j\rangle = \sum_{m=1}^{M} \mathcal{F}(a_m)\widehat{P}_m\,|\psi\rangle \tag{5.25}$$

for any vector $|\psi\rangle$, where $\widehat{P}_m := \sum_{j=1}^{d(m)} |a_m, j\rangle\langle a_m, j|$ is the projection operator onto the subspace of eigenvectors with eigenvalue a_m (see Eq. (3.56)). This definition makes sense provided only that $\mathcal{F}(a_m)$ is well-defined (*i.e.*, not infinite) for each eigenvalue a_m. Equivalently, we can write

$$\mathcal{F}(\widehat{A}) := \sum_{m=1}^{M} \sum_{j=1}^{d(m)} \mathcal{F}(a_m) |a_m, j\rangle\langle a_m, j| = \sum_{m=1}^{M} \mathcal{F}(a_m) \widehat{P}_m \tag{5.26}$$

which agrees with the definition Eq. (5.22) when $\mathcal{F}$ is a polynomial function of $a \in \mathbb{R}$, and which is consistent with the spectral representation of the self-adjoint operator $\widehat{A}$ (see Eq. (3.60)).

Indeed, starting with the spectral representation $\widehat{A} = \sum_{m=1}^{M} a_m \widehat{P}_m$ and the orthogonality condition Eq. (3.57), $\widehat{P}_m \widehat{P}_n = \delta_{mn} \widehat{P}_m$, we see at once that, for example, $\widehat{A}^2 = \sum_{m=1}^{M} a_m^2 \widehat{P}_m$ and, more generally, for any complex-valued polynomial function Q of $\mathbb{R}$,

$$Q(\widehat{A}) = \sum_{m=1}^{M} Q(a_m) \widehat{P}_m. \tag{5.27}$$

This strongly motivates the definition

$$\mathcal{F}(\widehat{A}) := \sum_{m=1}^{M} \mathcal{F}(a_m) \widehat{P}_m \tag{5.28}$$

which reproduces Eq. (5.26). Note that if $\mathcal{F}(a)$ is real for all a then the operator $\mathcal{F}(\widehat{A})$ is self-adjoint [Exercise].

5.3 The Alternative Form of Rule 3

5.3.1 The Main Theorem

Rule 3 looks somewhat different from the analogous rule in the summary in Chapter 1 of elementary wave mechanics. However we have the following theorem (with the simplifying assumption that the eigenvalue spectrum of $\widehat{A}$ is discrete):

Theorem

1. The only possible result of a measurement of A is one of the eigenvalues of the operator $\widehat{A}$ that represents it.

2. If the state vector is $|\psi\rangle$ and a measurement of A is made, the probability that the result will be the particular eigenvalue a_n is

$$\operatorname{Prob}(A = a_n; |\psi\rangle) = \langle\psi| \widehat{P}_n |\psi\rangle \tag{5.29}$$

where $\widehat{P}_n := \sum_{j=1}^{d(n)} |a_n, j\rangle\langle a_n, j|$ is the projector onto the eigenspace of vectors with eigenvalue a_n.

Proof

(1) The problem we face is to convert statements about *expected results* (which is all Rule 3 provides) into statements of the probabilities of specific results of measurements. The first step is to observe that, for any statistical system, the expected value of any function $\mathcal{F}(A)$ of a variable A is

$$\langle\mathcal{F}(A)\rangle = \sum_{m=1}^{M} \mathcal{F}(\alpha_m) \operatorname{Prob}(A = \alpha_m) \tag{5.30}$$

where $\{\alpha_1, \alpha_2, \ldots, \alpha_M\}$ is the set (for simplicity, assumed to be discrete) of all possible values of A, and $\operatorname{Prob}(A = \alpha_m)$ is the probability attached to the specific value α_m. In particular, for each $r \in \mathbb{R}$, consider the function χ_r of the real numbers defined by (*cf.* Eq. (4.10))

$$\chi_r(t) := \begin{cases} 1, & \text{if } t = r \\ 0 & \text{otherwise.} \end{cases} \tag{5.31}$$

Thus $\chi_r(A)$ is the quantity that equals 1 if $A = r$ and is 0 otherwise. In the language of classical physics discussed in Section 4.2, it represents the proposition that the physical quantity A has the value r. The expression Eq. (5.30) then implies that $\operatorname{Prob}(A = r) = \langle\chi_r(A)\rangle$; which solves the problem of how to derive the probability that A takes on some specific value from the expected values of other physical quantities.

In the case of quantum theory (and expressed in terms of measurement results, rather than possessed values), Eq. (5.30) implies that, for any state

$|\psi\rangle$, the expected result of measuring $\mathcal{F}(A)$ is

$$\langle \mathcal{F}(A)\rangle_\psi = \sum_{m=1}^{M} \mathcal{F}(\alpha_m)\,\text{Prob}(A = \alpha_m;\, |\psi\rangle), \tag{5.32}$$

and hence the special observable $\chi_r(A)$ has the property that

$$\text{Prob}(A = r;\, |\psi\rangle) = \langle \chi_r(A)\rangle_\psi. \tag{5.33}$$

Then, by Rule 3

$$\langle \chi_r(A)\rangle_\psi = \langle\psi|\, \widehat{\chi_r(A)}\, |\psi\rangle = \langle\psi|\, \chi_r(\hat{A})\, |\psi\rangle, \tag{5.34}$$

where the second equality follows from the function-preserving quantisation condition. However,

$$\chi_r(\hat{A}) := \sum_{m=1}^{M} \chi_r(a_m)\hat{P}_m = \begin{cases} \hat{P}_m, & \text{if } a_m = r, \\ 0 & \text{otherwise.} \end{cases} \tag{5.35}$$

Thus Eq. (5.33–5.34) show that $\text{Prob}(A = r;\, |\psi\rangle) = 0$ unless r is one of the eigenvalues of $\hat{A}$, which proves the first part of the theorem.

(2) Suppose that r *is* some eigenvalue a_n. Then the right hand side of Eq. (5.34) is $\langle\psi|\, \hat{P}_n\, |\psi\rangle$, and so $\text{Prob}(A = a_n;\, |\psi\rangle) = \langle\psi|\, \hat{P}_n\, |\psi\rangle$, as claimed.
QED

Comments

1. If the eigenvalue a_n is non-degenerate, then Eq. (5.29) gives the familiar result that the probability of obtaining a_n is $|\langle a_n|\psi\rangle|^2$.

2. The function-preserving condition $\widehat{\mathcal{F}(A)} = \mathcal{F}(\hat{A})$ plays a central part in proving this theorem. Indeed, this is one of the main reasons for being reluctant to give up the condition.

3. The rule $\langle A\rangle_\psi = \langle\psi|\, \hat{A}\, |\psi\rangle$ is entirely equivalent to the pair of rules

   (i) measurement results are always eigenvalues;

   (ii) $\text{Prob}(A = a_n;\, |\psi\rangle) = \langle\psi|\, \hat{P}_n\, |\psi\rangle$.

We have just shown that the rule $\langle A\rangle_\psi = \langle\psi|\,\widehat{A}\,|\psi\rangle$ implies (i) and (ii). Conversely, if we assume this pair of rules, then the usual definition of an expected result gives

$$
\begin{aligned}
\langle A\rangle_\psi &= \sum_{m=1}^{M} a_m \operatorname{Prob}(A = a_m;\psi) = \sum_{m=1}^{M} a_m \langle\psi|\,\widehat{P}_m\,|\psi\rangle \\
&= \sum_{m=1}^{M} \langle\psi|\,\widehat{A}\widehat{P}_m\,|\psi\rangle = \langle\psi|\,\widehat{A}\,|\psi\rangle
\end{aligned} \tag{5.36}
$$

as required. Note that, in deriving this, we have used the result $\widehat{A}\widehat{P}_m = a_m\widehat{P}_m$, and the resolution of the identity $\widehat{\mathbb{1}} = \sum_{m=1}^{M}\widehat{P}_m$.

4. The normalisation condition $\langle\psi|\psi\rangle = 1$ and the resolution of the identity $\widehat{\mathbb{1}} = \sum_{m=1}^{M}\widehat{P}_m$ imply that

$$
1 = \langle\psi|\psi\rangle = \sum_{m=1}^{M} \langle\psi|\,\widehat{P}_m\,|\psi\rangle \tag{5.37}
$$

so that, as required,

$$
\sum_{m=1}^{M} \operatorname{Prob}(A = a_m;\,|\psi\rangle) = 1. \tag{5.38}
$$

This result is essential for the internal consistency of the probabilistic rules of quantum theory, and it reveals the deep physical significance of the (mathematically, very non-trivial) expansion theorem Eq. (3.53–3.60). Thus the generalised Pythagorean theorem is seen to lie at the very heart of the formalism of quantum theory.

5. The second part of the theorem has an important implication. Suppose that the state vector is itself an eigenvector $|a_m\rangle$ of $\widehat{A}$, and that the observable A is then measured. According to Eq. (5.29), the relation $\langle a_n|a_m\rangle = \delta_{nm}$ implies that the probability of obtaining the result a_m is 1, and the probability of obtaining any other number is 0. Thus a measurement of A is guaranteed to yield the result a_m—a situation in which it arguably *is* meaningful to say that the system 'possesses' this value for A.

However, note that if B is another observable with $[\,\widehat{A},\widehat{B}\,] \neq 0$ then, in general, the state $|a_m\rangle$ will be a linear combination of eigenvectors of $\widehat{B}$.

Thus, even though the observable A has a 'value' in this state, this will not be the case for any other observable whose operator representative fails to commute with $\widehat{A}$ (see Section 6.3 for a general discussion of the significance of commuting and non-commuting pairs of operators).

6. Projection operators play a central role in both the technical and the conceptual foundations of quantum theory. We recall from Section 3.2 that an operator $\widehat{P}$ is a projection operator if and only if it satisfies $\widehat{P} = \widehat{P}^\dagger$ and $\widehat{P}^2 = \widehat{P}$. A typical example is $\widehat{P}_m := \sum_{j=1}^{d(m)} |a_m, j\rangle\langle a_m, j|$ which projects onto the subspace of eigenvectors of a self-adjoint operator $\widehat{A}$ with eigenvalue a_m.

The equation $\widehat{P}^2 = \widehat{P}$ implies that a projection operator $\widehat{P}$ has just the two eigenvalues 0 and 1, and hence, according to the theorem above, corresponds to an observable that can have only two values. For example, the operator $\widehat{P}_m$ represents the observable whose numerical value is defined operationally to be 1 if A is measured and the result a_m obtained, and is equal to 0 otherwise. Thus $\widehat{P}_m$ is associated with the *proposition* "if A is measured, the result obtained will be a_m". In a more realist reading, this proposition becomes "A *has* the value a_m". In classical physics, statements of this type are either true or false: a characteristic of normal propositional logic. As we shall see in our discussion of quantum logic in Section 9.2, the situation in quantum theory is potentially more complex.

5.3.2 A Simple Example

It may be helpful to illustrate the theorem above with the help of a simple example in quantum theory on a low-dimensional vector space.

Problem

The spin degrees of freedom of a ρ-meson can be represented on the vector space $\mathbb{C}^3$ by the following matrices

$$\widehat{S}_x = \frac{\hbar}{\sqrt{2}}\begin{pmatrix} 0 & 1 & 0 \\ 1 & 0 & 1 \\ 0 & 1 & 0 \end{pmatrix}, \quad \widehat{S}_y = \frac{\hbar}{\sqrt{2}}\begin{pmatrix} 0 & -i & 0 \\ i & 0 & -i \\ 0 & i & 0 \end{pmatrix}, \quad \widehat{S}_z = \hbar\begin{pmatrix} 1 & 0 & 0 \\ 0 & 0 & 0 \\ 0 & 0 & -1 \end{pmatrix}$$

(i) If a measurement is made of S_z what are the possible results?

(ii) If the state vector is $|\phi\rangle = \begin{pmatrix} 1 \\ i \\ -2 \end{pmatrix}$ what are the probabilities of obtaining the various results?

(iii) Check that these probabilities add up to one.

Answer

(i) The possible results of a measurement of S_z are the eigenvalues of the matrix $\widehat{S}_z$. Since this is a diagonal matrix, the eigenvalues are just the diagonal elements: $\hbar, 0, -\hbar$.

(ii) We shall need the normalised eigenvectors of this operator in order to compute the probabilities of obtaining the three possible results. By inspection, the eigenvectors corresponding to these eigenvalues $\hbar$, 0 and $-\hbar$ are

$$|+\hbar\rangle = \begin{pmatrix} 1 \\ 0 \\ 0 \end{pmatrix}, \quad |0\rangle = \begin{pmatrix} 0 \\ 1 \\ 0 \end{pmatrix}, \quad |-\hbar\rangle = \begin{pmatrix} 0 \\ 0 \\ 1 \end{pmatrix}$$

respectively. They are all normalised to one.

The next step is to find a normalised form of the state vector $|\phi\rangle = \begin{pmatrix} 1 \\ i \\ -2 \end{pmatrix}$. The norm of this vector is

$$\| \, |\phi\rangle \|^2 = \langle\phi|\phi\rangle = (1, -i, -2) \begin{pmatrix} 1 \\ i \\ -2 \end{pmatrix} = 6$$

and hence a normalised state vector is

$$|\psi\rangle := \frac{1}{\sqrt{6}} \begin{pmatrix} 1 \\ i \\ -2 \end{pmatrix}.$$

Thus the desired probabilities are:

$$(a) \quad \mathrm{Prob}(S_z = +\hbar; |\psi\rangle) = |\langle +\hbar|\psi\rangle|^2 = \frac{1}{6} \left| (1, 0, 0) \begin{pmatrix} 1 \\ i \\ -2 \end{pmatrix} \right|^2 = \frac{1}{6},$$

$$(b)\quad \text{Prob}(S_z = 0;\ |\psi\rangle) = |\langle 0|\psi\rangle|^2 = \frac{1}{6}\left|(0,\ 1,\ 0)\begin{pmatrix}1\\ i\\ -2\end{pmatrix}\right|^2 = \frac{1}{6},$$

$$(c)\quad \text{Prob}(S_z = -\hbar;\ |\psi\rangle) = |\langle -\hbar|\psi\rangle|^2 = \frac{1}{6}\left|(0,\ 0,\ 1)\begin{pmatrix}1\\ i\\ -2\end{pmatrix}\right|^2 = \frac{2}{3}.$$

(iii) The sum of the probabilities is $\frac{1}{6} + \frac{1}{6} + \frac{2}{3} = 1$, as required.

Chapter 6

TECHNICAL DEVELOPMENTS

6.1 Mixed States and Density Matrices

6.1.1 The Main Ideas

In this chapter we shall discuss various technical developments of the quantum rules presented in the previous chapter. We consider first how to generalise these rules to include the case of a mixed state $\rho := (\vec{\psi}_1, \vec{\psi}_2, \ldots, \vec{\psi}_D; w_1, w_2, \ldots, w_D)$, where the 'classical' probability that the vector state is $\vec{\psi}_d$ is w_d, with $0 < w_d \leq 1$ and $\sum_{d=1}^{D} w_d = 1$.

The central rule for a vector state $|\psi\rangle$ is $\mathrm{Prob}(A = a_n;\, |\psi\rangle) = \langle\psi|\, \widehat{P}_n\, |\psi\rangle$, and hence the corresponding rule for a mixed state is simply

$$\mathrm{Prob}(A = a_n; \rho) = \sum_{d=1}^{D} w_d\, \mathrm{Prob}(A = a_n;\, |\psi_d\rangle) = \sum_{d=1}^{D} w_d\, \langle\psi_d|\, \widehat{P}_n\, |\psi_d\rangle \quad (6.1)$$

on the assumption that the quantum-mechanical and classical probabilities are independent of each other.

Our main task is to express Eq. (6.1) in a more succinct form. To this end, let $\widehat{B}$ be any operator, and let $\{\, |e_1\rangle,\, |e_2\rangle, \ldots,\, |e_N\rangle\}$ be an orthonormal basis of $\mathcal{H}$. Then the *trace* of $\widehat{B}$ is defined by

$$\operatorname{tr}\widehat{B} := \sum_{i=1}^{N} \langle e_i|\,\widehat{B}\,|e_i\rangle \tag{6.2}$$

which can be shown [Exercise] to be independent of the basis chosen.[1] This is clearly a natural generalisation of the trace of a matrix.

Now let $|\psi\rangle$ be any vector in $\mathcal{H}$. Then

$$\begin{aligned}\langle\psi|\,\widehat{B}\,|\psi\rangle &= \sum_{i=1}^{N}\langle\psi|\,\widehat{B}\,|e_i\rangle\langle e_i|\psi\rangle \equiv \sum_{i=1}^{N}\langle e_i|\psi\rangle\langle\psi|\,\widehat{B}\,|e_i\rangle \\ &= \sum_{i=1}^{N}\langle e_i|\,\widehat{P}_\psi\widehat{B}\,|e_i\rangle = \operatorname{tr}(\widehat{P}_\psi\widehat{B}) \end{aligned}\tag{6.3}$$

where $\widehat{P}_\psi := |\psi\rangle\langle\psi|$. Therefore, in terms of the trace operation, the basic probabilistic rule Eq. (5.29) can be rewritten as

$$\operatorname{Prob}(A = a_n;\, |\psi\rangle) = \operatorname{tr}(\widehat{P}_\psi\widehat{P}_n), \tag{6.4}$$

while the expression for the expected result is

$$\langle A\rangle_\psi = \operatorname{tr}(\widehat{P}_\psi\widehat{A}). \tag{6.5}$$

Now define the *mixed-state operator* $\widehat{\rho}$ by

$$\widehat{\rho} := \sum_{d=1}^{D} w_d\,\widehat{P}_{|\psi_d\rangle} \tag{6.6}$$

where, we recall, w_d is the classical probability that the vector state is $|\psi_d\rangle$. Then, because $\operatorname{tr}\widehat{B}$ is a linear function of the operator $\widehat{B}$, Eq. (6.1) becomes

$$\operatorname{Prob}(A = a_n;\rho) = \sum_{d=1}^{D} w_d \operatorname{tr}(\widehat{P}_{|\psi_d\rangle}\widehat{P}_n) = \operatorname{tr}\Big(\sum_{d=1}^{D} w_d\widehat{P}_{|\psi_d\rangle}\widehat{P}_n\Big)$$

so that

$$\boxed{\operatorname{Prob}(A = a_n;\rho) = \operatorname{tr}(\widehat{\rho}\,\widehat{P}_n).} \tag{6.7}$$

[1] In the infinite-dimensional case it is necessary to restrict attention to those operators whose trace actually exists, *i.e.*, for which the infinite sum converges. These are called *trace-class* operators.

The corresponding expression for the expected result is

$$\boxed{\langle A\rangle_\rho = \operatorname{tr}(\widehat{\rho}\,\widehat{A}).} \tag{6.8}$$

Note that these expressions reproduce the usual results for a pure vector state $|\psi\rangle$ if the special density matrix $\widehat{\rho} = \widehat{P}_\psi = |\psi\rangle\langle\psi|$ is used.

Comments

1. The operator $\widehat{\rho}$ defined in Eq. (6.6) has the following properties:

   (i) $\widehat{\rho} = \widehat{\rho}^\dagger$;

   (ii) $\widehat{\rho}$ is a *positive, semi-definite* operator, *i.e.*, $\langle\psi|\,\widehat{\rho}\,|\psi\rangle \geq 0$ for all vectors $|\psi\rangle$;

   (iii) $\operatorname{tr}\widehat{\rho} = 1$.

In general, any operator satisfying these three conditions is called a *density matrix*.

It is straightforward to see that if $\widehat{\rho}$ is *any* such operator (*i.e.*, not necessarily of the form Eq. (6.6)), then Eq. (6.7) satisfies the basic requirements Eq. (2.1–2.2) for a probability assignment. Thus we appear to have extended the type of mathematical object that can give probabilities. However, since $\widehat{\rho}$ is a self-adjoint operator, the spectral theorem can be applied, and this shows that $\widehat{\rho}$ can, in fact, be written in the form Eq. (6.6) where the real numbers $w_1, w_2, \ldots, w_D$ are now the eigenvalues of $\widehat{\rho}$. Thus nothing new is really achieved by considering general density matrices. However, note that:

- Some of these eigenvalues may be equal to each other and, in that case, there is no unique way of choosing the vectors that span the degeneracy subspaces for the distinct eigenvalues.

- If the spectral theorem is used on a given density matrix $\widehat{\rho}$, the vectors appearing in the decomposition Eq. (6.6) will be pairwise orthogonal, but this property is not necessary for a meaningful notion of classically weighted sums of states.

These two features reflect the important fact that a given density matrix does not admit a unique decomposition of the type Eq. (6.6). For example, for a spinning electron, the operators that project onto the eigenstates of $\widehat{S}_z$ with eigenvalues $\frac{\hbar}{2}$ and $\frac{-\hbar}{2}$ are respectively

$$\widehat{P}_{S_z=\hbar/2} = \begin{pmatrix} 1 & 0 \\ 0 & 0 \end{pmatrix}, \quad \widehat{P}_{S_z=-\hbar/2} = \begin{pmatrix} 0 & 0 \\ 0 & 1 \end{pmatrix} \tag{6.9}$$

whereas the operators that project onto the eigenstates of $\widehat{S}_x$ are

$$\widehat{P}_{S_x=\hbar/2} = \frac{1}{2}\begin{pmatrix} 1 & 1 \\ 1 & 1 \end{pmatrix}, \quad \widehat{P}_{S_x=-\hbar/2} = \frac{1}{2}\begin{pmatrix} 1 & -1 \\ -1 & 1 \end{pmatrix}. \tag{6.10}$$

Then, for example, the density matrix $\widehat{\rho} := \frac{1}{2}\left(\begin{smallmatrix} 1 & 0 \\ 0 & 1 \end{smallmatrix}\right)$ can be written as the sum $\frac{1}{2}\widehat{P}_{S_z=\hbar/2} + \frac{1}{2}\widehat{P}_{S_z=-\hbar/2}$, which suggests an interpretation in which there is an equal probability of the quantum state being the eigenstate $|S_z = \hbar/2\rangle$ or the eigenstate $|S_z = -\hbar/2\rangle$. One can imagine producing such a density matrix by randomly mixing the outputs of a pair of Stern–Gerlach devices[2] that are oriented along the z-axis, and whose output beams are filtered so that one passes only spin-up electrons, and the other passes only spin-down electrons.

On the other hand, $\widehat{\rho}$ can also be written as the sum $\frac{1}{2}\widehat{P}_{S_x=\hbar/2} + \frac{1}{2}\widehat{P}_{S_x=-\hbar/2}$, which suggests an interpretation as a stochastic mixture of eigenstates of S_x. But this is not the same as the interpretation using S_z. This non-uniqueness of the decomposition makes it difficult to sustain any simple interpretation of a density matrix as a mathematical measure of our ignorance of which pure states should be used to describe the system.

2. Let $\{\widehat{\rho}_1, \widehat{\rho}_2, \ldots, \widehat{\rho}_K\}$ be any set of density matrices, and let $\{r_1, r_2, \ldots, r_K\}$ be any set of positive real numbers satisfying $\sum_{k=1}^{K} r_k = 1$. Then it is easy to see that the *affine sum* $\sum_{k=1}^{K} r_k \widehat{\rho}_k$ also satisfies the requirements for a density matrix. This means that, regarded as a subset of the vector space of all bounded operators on the Hilbert space $\mathcal{H}$, the set of density matrices is *convex*. The special density matrices $\widehat{P}_\psi := |\psi\rangle\langle\psi|$ are

[2]Real Stern–Gerlach devices do not act on a beam of electrons but rather on a beam of silver atoms whose net angular momentum is $\frac{1}{2}\hbar$. However, that is not important here.

the only ones that cannot be decomposed into an affine sum in this way; they can be viewed geometrically therefore as lying on the boundary of the convex set. States of this type are called *pure*; the rest are *mixed*. This particular view of the space of quantum states plays an important role in investigations of the most general mathematical structure that might be used to describe quantum systems.

3. From a practical perspective, one of the most important examples of a mixed state is the *thermal state* that is appropriate for describing a quantum statistical system at a temperature T. The corresponding density matrix is

$$\widehat{\rho}_T := \frac{e^{-\widehat{H}/kT}}{Z(T)} \tag{6.11}$$

where

$$Z(T) := \operatorname{tr}(e^{-\widehat{H}/kT}) \tag{6.12}$$

is the *partition function*, and k is Boltzmann's constant. Using the spectral theorem for the Hamiltonian $\widehat{H}$ (and assuming, for simplicity, that the energy eigenvalues $\{E_1, E_2, \ldots, E_M\}$ are non-degenerate and discrete), $\widehat{\rho}_T$ can be written as

$$\widehat{\rho}_T := \sum_{m=1}^{M} \frac{e^{-E_m/kT}}{Z(T)} \, |E_m\rangle\langle E_m| \tag{6.13}$$

which reflects a classical uncertainty, induced by thermal fluctuations, of which energy state E_m the system is in. According to Eq. (6.13), the probability that it is $|E_m\rangle$ is $e^{-E_m/kT}/Z(T)$, where the partition function $Z(T)$ can be computed from Eq. (6.12) as

$$Z(T) = \sum_{m=1}^{M} e^{-E_m/kT} \tag{6.14}$$

on using the basis set $\{\,|E_1\rangle, |E_2\rangle, \ldots\,\}$ to calculate the trace.

6.1.2 A Simple Example

Like much else in quantum theory, the ideas of density matrices can be usefully illustrated with the aid of a simple model system on the vector space $\mathbb{C}^2$.

Problem

The spin state of an electron is represented on $\mathbb{C}^2$ (in the basis formed by the eigenstates of $\widehat{S}_z$) by the density matrix $\rho := \begin{pmatrix} a & 0 \\ 0 & b \end{pmatrix}$ where a and b are real numbers with $a \geq 0$, $b \geq 0$ and $a + b = 1$.

(i) If a measurement is made of the spin S_x what is the probability that the result obtained will be (a) $\frac{\hbar}{2}$; (b) $\frac{-\hbar}{2}$?

(ii) Use these results to compute the expected result, and check that the answer agrees with the result calculated directly from the density matrix formalism using Eq. (6.8).

Answer

(i) We know in general from Eq. (6.7) that $\text{Prob}(A = a_n; \rho) = \text{tr}\,(\hat{\rho}\,\widehat{P}_{A=a_n})$, where $\widehat{P}_{A=a_n}$ is the projection operator onto the subspace of eigenvectors of $\widehat{A}$ with eigenvalue equal to a_n. In our case, the eigenvalues of the spin operator $S_x = \frac{\hbar}{2}\begin{pmatrix} 0 & 1 \\ 1 & 0 \end{pmatrix}$ are non-degenerate, and equal to $\frac{1}{2}\hbar$ and $-\frac{1}{2}\hbar$. A corresponding set of normalised eigenvectors is $|\rightarrow\rangle := \frac{1}{\sqrt{2}}\begin{pmatrix} 1 \\ 1 \end{pmatrix}$ and $|\leftarrow\rangle := \frac{1}{\sqrt{2}}\begin{pmatrix} 1 \\ -1 \end{pmatrix}$ (see Problem 9 at the end of the book for the details of the required calculations). Thus the crucial projection operators are $\widehat{P}_{S_x=\hbar/2} = P_{|\rightarrow\rangle}$ and $\widehat{P}_{S_x=-\hbar/2} = P_{|\leftarrow\rangle}$ (see below).

The statement that an operator on a vector space is represented by a specific matrix means that this is true with respect to some particular basis set. In the present case, we are making the standard choice of the basis set given by the eigenvectors $|e_1\rangle := \begin{pmatrix} 1 \\ 0 \end{pmatrix}$ and $|e_2\rangle := \begin{pmatrix} 0 \\ 1 \end{pmatrix}$ of the S_z operator. Thus the projection operator onto the state $|\rightarrow\rangle$ has the matrix elements

$$(\widehat{P}_{|\rightarrow\rangle})_{ij} := \langle e_i|\,\widehat{P}_{|\rightarrow\rangle}\,|e_j\rangle = \langle e_i|\rightarrow\rangle\langle\leftarrow|e_j\rangle$$

for $i, j = 1, 2$. This is just the 2×2 matrix

$$\widehat{P}_{|\rightarrow\rangle} = \frac{1}{2}\begin{pmatrix} 1 & 1 \\ 1 & 1 \end{pmatrix},$$

and similarly

$$\widehat{P}_{|\leftarrow\rangle} = \frac{1}{2}\begin{pmatrix} 1 & -1 \\ -1 & 1 \end{pmatrix}.$$

Then

$$\widehat{\rho}\,\widehat{P}_{|\rightarrow\rangle} = \frac{1}{2}\begin{pmatrix} a & 0 \\ 0 & b \end{pmatrix}\begin{pmatrix} 1 & 1 \\ 1 & 1 \end{pmatrix} = \frac{1}{2}\begin{pmatrix} a & a \\ b & b \end{pmatrix}$$

whose trace is equal to $\frac{1}{2}(a+b) = \frac{1}{2}$. Similarly

$$\widehat{\rho}\,\widehat{P}_{|\leftarrow\rangle} = \frac{1}{2}\begin{pmatrix} a & 0 \\ 0 & b \end{pmatrix}\begin{pmatrix} 1 & -1 \\ -1 & 1 \end{pmatrix} = \frac{1}{2}\begin{pmatrix} a & -a \\ -b & b \end{pmatrix}$$

whose trace is also $\frac{1}{2}(a+b) = \frac{1}{2}$. Thus

$$\text{Prob}(S_x = \tfrac{\hbar}{2}; \rho) = \frac{1}{2} = \text{Prob}(S_x = -\tfrac{\hbar}{2}; \rho).$$

(ii) The expected result of a measurement of S_x in the state ρ is clearly $\frac{1}{2}\frac{\hbar}{2} + (-\frac{1}{2})\frac{\hbar}{2} = 0$. On the other hand, the general expression for the expected result of a measurement of an observable A in a mixed state $\widehat{\rho}$ is $\langle A\rangle_\rho = \text{tr}\,(\widehat{\rho}\,\widehat{A})$. In our case,

$$\widehat{A} = \widehat{S}_x = \frac{\hbar}{2}\begin{pmatrix} 0 & 1 \\ 1 & 0 \end{pmatrix}$$

and hence

$$\langle S_x\rangle_\rho = \frac{\hbar}{2}\,\text{tr}\left\{\begin{pmatrix} a & 0 \\ 0 & b \end{pmatrix}\begin{pmatrix} 0 & 1 \\ 1 & 0 \end{pmatrix}\right\} = \frac{\hbar}{2}\,\text{tr}\begin{pmatrix} 0 & a \\ b & 0 \end{pmatrix} = 0,$$

which is consistent with the previous calculation.

6.2 Operators With a Continuous Spectrum

As remarked in Chapter 3, in an infinite-dimensional Hilbert space there are self-adjoint operators whose eigenvalues form continuous ranges of numbers. Two examples in wave mechanics are $\widehat{x}$ and $-i\hbar\frac{d}{dx}$, both of whose eigenvalues fill the entire real line. This causes various problems. For example, although the original form of Rule 3 is unchanged, it is no longer meaningful to talk about the probability of finding a particular eigenvalue. Rather, one must speak of the probability that the result of measuring A will lie in a certain *range*. This is related mathematically to the fact, mentioned in Chapter 3, that the corresponding eigenvectors are not normalisable, and hence cannot be used as *bona fide* state vectors. This is not unreasonable physically, since a real number can never be measured with absolute accuracy: in practice, the result is only known to lie in some interval. In particular, these non-normalisable eigenstates cannot be prepared by any physical process.

The full mathematical theory of continuous eigenvalues is quite difficult, and can be developed in various ways. One approach, due originally to von Neumann, is based on an extension of the spectral theorem to include any self-adjoint operator $\widehat{A}$ whose spectrum is continuous (von Neumann 1971). This theorem shows the existence of a one-parameter family of projection operators $\widehat{E}_a$, $a \in \mathbb{R}$, such that, for any pair of vectors $\vec{\psi}, \vec{\phi}$ in $\mathcal{H}$, the matrix element $\langle \vec{\psi}, \widehat{A}\vec{\phi} \rangle$ can be written as the integral

$$\langle \vec{\psi}, \widehat{A}\vec{\phi} \rangle = \int_{-\infty}^{\infty} a \, d_a \langle \vec{\psi}, \widehat{E}_a \vec{\phi} \rangle. \tag{6.15}$$

If $\langle \vec{\psi}, \widehat{E}_a \vec{\phi} \rangle$ is a differentiable function of $a \in \mathbb{R}$, the expression $d_a \langle \vec{\psi}, \widehat{E}_a \vec{\phi} \rangle$ can be understood as $\frac{d}{da} \langle \vec{\psi}, \widehat{E}_a \vec{\phi} \rangle \, da$, and the right hand side of Eq. (6.15) is a normal integral. If $\langle \vec{\psi}, \widehat{E}_a \vec{\phi} \rangle$ is not differentiable, Eq. (6.15) must be interpreted as a Stieltjes integral.

With due care, for each interval[3] $[b, c]$ of the real numbers (with $b < c$), a projection operator can be defined by

$$\widehat{E}_{[b,c]} := \int_b^c d\widehat{E}_a. \tag{6.16}$$

This projects onto the subspace of states for which the eigenvalues a of $\widehat{A}$ lie in the interval $[b, c]$. Physically, the probability that a measurement of A will yield a result lying in this interval is

$$\text{Prob}(A \in [\, b, c]; |\psi\rangle) = \langle \psi | \, \widehat{E}_{[b,c]} \, | \psi \rangle. \tag{6.17}$$

Note however that this requires a strict inequality $b < c$: there is *no* projector onto an eigenspace corresponding to a *single* eigenvalue. In particular, there are no vectors $|a\rangle$ in $\mathcal{H}$ which are eigenvectors of $\widehat{A}$ with $\widehat{A}\,|a\rangle = a\,|a\rangle$.

[3]The notation $[b, c]$ means the set of all real numbers $a \in \mathbb{R}$ such that $b \leq a \leq c$.

A somewhat different approach to operators with continuous spectra was adopted by Dirac who simply assumed that eigenvectors $|a\rangle$ must exist in some sense. The full mathematical implementation of this idea involves what is known as a *rigged Hilbert space*, or a *Gel'fand triple* $\{\Phi, \mathcal{H}, \Phi'\}$, in which the Hilbert space $\mathcal{H}$ appears as a linear subspace of a vector space Φ' which is not itself a Hilbert space. Eigenvectors associated with a continuous spectrum are *bona fide* vectors in Φ' but are not members of the subspace $\mathcal{H}$. The space Φ' is constructed as a type of dual (see Section 2.5.2) of another vector space Φ which is a subspace of $\mathcal{H}$. Thus we have $\Phi \subset \mathcal{H} \subset \Phi'$.

The full treatment of this case is quite subtle, but most physicists get along very well by pretending that vectors $|a\rangle$ exist, and then manipulating them according to the following heuristic rules laid out by Dirac (in what follows we will assume for simplicity that $\widehat{A}$ is multiplicity-free, *i.e.*, there are no degenerate eigenvalues):

(i) If $\widehat{A}\,|a\rangle = a\,|a\rangle$ for a in some continuous range $\alpha \leq a \leq \beta$ (α and β could be $-\infty$ and ∞ respectively), then the eigenvectors $|a\rangle$ can be orthonormalised with respect to the Dirac delta function (which, of course, is not a 'proper' function in the rigorous sense) so that

$$\langle a|b\rangle = \delta(a-b) \tag{6.18}$$

where, for any function f, $\int_{-\infty}^{\infty} f(a)\delta(a-b)\,da = f(b)$. This should be compared with the discrete results $\langle a_n|a_m\rangle = \delta_{nm}$ and $\sum_n f_n\delta_{nm} = f_m$.

(ii) Any vector in $\mathcal{H}$ can be expanded in terms of these generalised eigenvectors as

$$|\psi\rangle = \int_\alpha^\beta \psi(a)\,|a\rangle\,da \tag{6.19}$$

where the role of the expansion coefficients is played by the function ψ defined by

$$\psi(a) := \langle a|\psi\rangle. \tag{6.20}$$

This should be compared with the discrete result $|\psi\rangle = \sum_m \psi_m\,|a_m\rangle$, where the expansion coefficients are $\psi_m := \langle a_m|\psi\rangle$. There are similar analogues of the resolution of the identity Eq. (3.23), and the spectral representation Eq. (3.55).

(iii) If $|\psi\rangle = \int_\alpha^\beta \psi(a)\,|a\rangle\,da$ and $|\phi\rangle = \int_\alpha^\beta \phi(b)\,|b\rangle$ then, heuristically,

$$\begin{aligned}\langle\psi|\phi\rangle &= \int_\alpha^\beta\int_\alpha^\beta \psi^*(a)\phi(b)\langle a|b\rangle\,da\,db = \int_\alpha^\beta\int_\alpha^\beta \psi^*(a)\phi(b)\delta(a-b)\,da\,db \\ &= \int_\alpha^\beta \psi^*(a)\phi(a)\,da \end{aligned} \tag{6.21}$$

which should be compared with the discrete result $\langle\psi|\phi\rangle = \sum_n \psi_n^*\phi_n$. In particular,

$$\|\vec{\psi}\|^2 = \int_\alpha^\beta |\psi(a)|^2\,da \tag{6.22}$$

and the 'wave packet' in Eq. (6.19) will correspond to a physically-realisable state provided the function $\psi(a)$ is chosen so that this integral is finite.

(iv) Ignoring degeneracy, the key result Eq. (5.29) becomes the probability that a measurement of the observable A will yield a result lying in the range $[b, c]$ is

$$\text{Prob}(A \in [b,c];\, |\psi\rangle) = \int_b^c |\langle a|\psi\rangle|^2\,da. \tag{6.23}$$

Note that the continuum eigenvectors $|a\rangle$ are connected with von Neumann's spectral projectors $\widehat{E}_a$ by the heuristic expression

$$d\widehat{E}_a = |a\rangle\langle a|\,da \tag{6.24}$$

which reconciles the heuristic equation Eq. (6.23) with the mathematically-rigorous expression Eq. (6.17). If the spectrum of $\widehat{A}$ contains both continuous and discrete parts then both sums and integrals are needed, but we shall not go into details here.

6.3 Compatible Observables

In debates on the meaning of quantum theory one frequently encounters references to the question of whether two physical quantities A and B are *compatible* with each other: something that is usually regarded as being synonymous with the requirement that they can be measured simultaneously or, in a more realist interpretation, that they can have values simultaneously.

Such a requirement is trivial in classical mechanics; indeed, it is inherent in the association of each physical quantity A with a real-valued function f_A on the space $\mathcal{S}$ of states, such that the value of A in the state $s \in \mathcal{S}$ is $f_A(s)$. The situation in quantum theory is quite different since, as we shall see, the incompatibility of many pairs of observables is, arguably, a direct consequence of the formalism itself. Of course, such a statement requires a clear definition of what is meant by a 'simultaneous measurement'. Any measurement takes a finite, albeit small, amount of time. Are measurements of two observables A and B to be regarded as simultaneous if those time intervals only partially overlap? A more promising definition might be that A and B are simultaneously measurable if a measurement of A followed immediately by a measurement of B yields the same results as when the time order of the A and B measurements is reversed. In any case, one might worry about the implications of the special theory of relativity in which the notion of simultaneity is reference-frame dependent.

In this section we shall concentrate mainly on the technical aspects associated with the idea of compatibility. Of particular interest are the implications of the fact that certain types of measurement clearly *can* be regarded as being simultaneous without becoming entangled in conceptual issues. For example, consider the measurement of the x and y coordinates of an electron by allowing it to hit a phosphor screen in the x–y plane. If the phosphor points are numbered, a single observation of which point is hit (*i.e.*, its number) enables a reconstruction of the x and y coordinates of the point; as such, it constitutes a simultaneous measurement of both the x and the y observables of the electron. This is a particular case of the general situation in which two observables A and B are said to be *trivially compatible* if there is a third observable C and functions $\mathcal{F}$ and $\mathcal{G}$ such that $A = \mathcal{F}(C)$ and $B = \mathcal{G}(C)$. Thus a number can be ascribed to both A and B by a single measurement of C. The function-preserving quantisation condition Eq. (5.14) (which is assumed here) then implies that $\widehat{A} = \mathcal{F}(\widehat{C})$ and $\widehat{B} = \mathcal{G}(\widehat{C})$, and hence that

$$[\widehat{A}, \widehat{B}] = 0 \tag{6.25}$$

since a straightforward consequence of the spectral theorem is that any two functions of a self-adjoint operator commute with each other. [Exercise]

It is clearly of some importance to know when a pair of observables is trivially compatible in this sense; *i.e.*, when they are both functions of a third quantity. In this context, the crucial mathematical result is that the *converse* of the deduction of Eq. (6.25) is true in the following sense.

Theorem

Let $\widehat{A}$ and $\widehat{B}$ be a pair of self-adjoint operators with $[\,\widehat{A},\widehat{B}\,]=0$. Then there exists a third self-adjoint operator $\widehat{C}$, and functions $\mathcal{F},\mathcal{G}:\mathbb{R}\rightarrow\mathbb{R}$, such that $\widehat{A}=\mathcal{F}(\widehat{C})$ and $\widehat{B}=\mathcal{G}(\widehat{C})$.

Proof

A sketch of the proof (in the case where both operators have a discrete spectrum) is as follows.

Write the spectral representations of $\widehat{A}$ and $\widehat{B}$ as

$$\begin{aligned}\widehat{A} &= \sum_{i=1}^{M_A} a_i\widehat{P}_i \\ \widehat{B} &= \sum_{j=1}^{M_B} b_j\widehat{Q}_j,\end{aligned} \tag{6.26}$$

where $\{a_1,a_2,\ldots,a_{M_A}\}$ and $\{b_1,b_2,\ldots,b_{M_B}\}$ are the eigenvalues of $\widehat{A}$ and $\widehat{B}$ respectively. Then it is straightforward to show that $[\widehat{A},\widehat{B}]=0$ implies (i) $[\,\widehat{A},\widehat{Q}_j]=0$, (ii) $[\,\widehat{P}_i,\widehat{B}\,]=0$, and (iii) $[\widehat{P}_i,\widehat{Q}_j]=0$, for all i and j (see problem 16 at the end of the book).

Since $\widehat{P}_i\widehat{Q}_j=\widehat{Q}_j\widehat{P}_i$ it follows that $(\widehat{P}_i\widehat{Q}_j)^\dagger=\widehat{Q}_j\widehat{P}_i=\widehat{P}_i\widehat{Q}_j$, *i.e.*, the operator $\widehat{P}_i\widehat{Q}_j$ is self-adjoint. Furthermore,

$$(\widehat{P}_i\widehat{Q}_j)^2=\widehat{P}_i\widehat{Q}_j\widehat{P}_i\widehat{Q}_j=\widehat{P}_i\widehat{Q}_j\widehat{Q}_j\widehat{P}_i=\widehat{P}_i\widehat{Q}_j\widehat{P}_i=\widehat{P}_i\widehat{P}_i\widehat{Q}_j=\widehat{P}_i\widehat{Q}_j \tag{6.27}$$

and hence, for each i and j, $\widehat{P}_i\widehat{Q}_j$ is a projection operator. Of course, some of these may vanish, so let us define $S(\widehat{A},\widehat{B})$ to be the subset of pairs i and j for which $\widehat{P}_i\widehat{Q}_j\neq 0$. Then, if $\{c_{ij}\mid (i,j)\in S(\widehat{A},\widehat{B})\}$, is any collection of real numbers, an associated self-adjoint operator $\widehat{C}$ can be defined by

$$\widehat{C}:=\sum_{i,j\in S(\widehat{A},\widehat{B})} c_{ij}\widehat{P}_i\widehat{Q}_j. \tag{6.28}$$

Note that, since $\sum_{j=1}^{M_B}\widehat{Q}_j=\widehat{1}$, for each i there exists at least one value of j such that $\widehat{P}_i\widehat{Q}_j\neq 0$.

Now choose a set of such numbers $\{c_{ij} \mid (i,j) \in S(\widehat{A}, \widehat{B})\}$ that are all non-zero and different from each other, and construct a function $\mathcal{F} : \mathbb{R} \to \mathbb{R}$ by first defining $\mathcal{F}$ on the set $\{c_{ij}\}$ by $\mathcal{F}(c_{ij}) := a_i$ for all $i, j \in S(\widehat{A}, \widehat{B})$, and then interpolating $\mathcal{F}$ between the elements of this set in any arbitrary way. Note that, since for each i there is at least one index j such that $c_{ij} \neq 0$, every a_i, $i = 1, 2, \ldots, M_A$ is in the range of $\mathcal{F}$. Then it follows from the definition Eq. (6.28) of $\widehat{C}$ that $\widehat{A} = \mathcal{F}(\widehat{C})$.

Similarly, we can define a function $\mathcal{G}$ so that, for all c_{ij} we have $\mathcal{G}(c_{ij}) = b_j$. This gives $\widehat{B} = \mathcal{G}(\widehat{C})$. **QED**

Further insight can be gained by looking in more detail at the projection operators $\widehat{P}_i\widehat{Q}_j$. In particular, if $|\psi\rangle$ is any vector in $\mathcal{H}$ we see that $\widehat{A}\widehat{P}_i\widehat{Q}_j\,|\psi\rangle = a_i\widehat{P}_i\widehat{Q}_j\,|\psi\rangle$ (since $A\widehat{P}_i = a_i\widehat{P}_i$) and $\widehat{B}\widehat{P}_i\widehat{Q}_j\,|\psi\rangle = \widehat{P}_i\widehat{B}\widehat{Q}_j\,|\psi\rangle = b_j\widehat{P}_i\widehat{Q}_j\,|\psi\rangle$ (since $[\widehat{B}, \widehat{P}_i] = 0$, and $\widehat{B}\widehat{Q}_j = b_j\widehat{Q}_j$). Thus vectors of the form $\widehat{P}_i\widehat{Q}_j\,|\psi\rangle$ are *simultaneous eigenvectors* of the operators $\widehat{A}$ and $\widehat{B}$ with eigenvalues a_i and b_j respectively.

The set of vectors $\{\widehat{P}_i\widehat{Q}_j\,|\psi\rangle$ for $|\psi\rangle \in \mathcal{H}\}$ is the eigenspace of the projection operator $\widehat{P}_i\widehat{Q}_j$ with eigenvalue 1, and we can find an orthonormal basis set for this subspace with vectors labelled $|a_i, b_j, k\rangle$, where $k = 1, 2, \ldots, d(i,j)$ is the degeneracy label for the pair of eigenvalues a_i and b_j (*i.e.*, $d(i,j)$ is the dimension of the eigenspace). Repeating this for all i and j gives a basis for the whole Hilbert space, in terms of which any vector $|\psi\rangle$ can be expanded as

$$|\psi\rangle = \sum_{i,j\in S(\widehat{A},\widehat{B})} \sum_{k=1}^{d(i,j)} \psi_{ijk}\,|a_i, b_j, k\rangle. \tag{6.29}$$

Note that, as far as the operator $\widehat{A}$ is concerned, the degenerate eigenstates associated with the eigenvalue a_i are partially labelled by the eigenvalues of $\widehat{B}$ and partially by the index k. This raises the question of whether k can be related to the eigenvalues of a third operator that commutes with both $\widehat{A}$ and $\widehat{B}$, and so on. The important answer is contained in the following theorem:

Theorem

On any Hilbert space $\mathcal{H}$ there are finite sets of self-adjoint operators $\widehat{A}_1, \widehat{A}_2, \ldots, \widehat{A}_r$ that commute in pairs and whose simultaneous eigenvectors form a basis set for $\mathcal{H}$ with no residual degeneracy. Any set of commuting self-adjoint operators with residual degeneracy can be made into one of these non-degenerate sets by the addition of a finite set of pair-wise commuting operators.

Comments

1. Any vector $|\psi\rangle \in \mathcal{H}$ can be expanded in terms of the associated simultaneous eigenvectors $|a_{1i_1}, a_{2i_2}, \ldots, a_{ri_r}\rangle$ (with $\widehat{A}_k\,|a_{1i_1}, a_{2i_2}, \ldots, a_{ri_r}\rangle = a_{ki_k}\,|a_{1i_1}, a_{2i_2}, \ldots, a_{ri_r}\rangle$) as

$$|\psi\rangle = \sum_{i_1,i_2,\ldots,i_r} \psi_{i_1,i_2,\ldots,i_r}\,|a_{1i_1}, a_{2i_2}, \ldots, a_{ri_r}\rangle \tag{6.30}$$

where the expansion coefficients are

$$\psi_{i_1,i_2,\ldots,i_r} = \langle a_{1i_1}, a_{2i_2}, \ldots, a_{ri_r}|\psi\rangle \tag{6.31}$$

and where the sum in Eq. (6.30) is over the eigenvalues of the operators $\widehat{A}_1, \widehat{A}_2, \ldots, \widehat{A}_r$.

2. The basic probabilistic rule Eq. (5.29) now becomes the statement that if the state vector is $|\psi\rangle$, and if a simultaneous measurement is made of the trivially compatible observables $A_1, A_2, \ldots, A_r$, the joint probability that the results obtained will be $a_{1i_1}, a_{2i_2}, \ldots, a_{ri_r}$ is

$$\text{Prob}(A_1 = a_{1i_1}, A_2 = a_{2i_2}, \ldots, A_r = a_{ri_r};\, |\psi\rangle) = |\langle a_{1i_1}, a_{2i_2}, \ldots, a_{ri_r}|\psi\rangle|^2. \tag{6.32}$$

There is an obvious extension of the expression to include mixed states.

3. Dirac called sets of operators with this property *complete commuting sets*. Note that it is meaningful to say that an individual quantum system described by the quantum state vector $|a_{1i_1}, a_{2i_2}, \ldots, a_{ri_r}\rangle$ *possesses* the simultaneous values $a_{1i_1}, a_{2i_2}, \ldots, a_{ri_r}$ for the observables $A_1, A_2, \ldots, A_r$.

4. A typical example is the state vectors, $|nlm\lambda\rangle$, of a hydrogen atom. These are simultaneous eigenvectors of the Hamiltonian $\widehat{H}$, the total angular momentum $\widehat{L}\cdot\widehat{L}$, the z-component $\widehat{L}_z$ of angular momentum, and the z-component $\widehat{S}_z$ of the spin operator of the electron. The eigenvalues are

$$\widehat{H}\,|nlm\lambda\rangle = E_n\,|nlm\lambda\rangle, \quad n = 1, 2, \ldots \tag{6.33}$$

$$\begin{aligned}
\widehat{L}\cdot\widehat{L}\,|nlm\lambda\rangle &= l(l+1)\hbar^2\,|nlm\lambda\rangle, \quad l=0,1,\ldots,n-1 && (6.34)\\
\widehat{L}_z\,|nlm\lambda\rangle &= m\hbar\,|nlm\lambda\rangle, \quad m=-l,-l+1,\ldots,l-1,l+1 && (6.35)\\
\widehat{S}_z\,|nlm\lambda\rangle &= \lambda\hbar\,|nlm\lambda\rangle, \quad \lambda=\pm\tfrac{1}{2}. && (6.36)
\end{aligned}$$

The main physical conclusion of all this mathematical discussion is that observables A and B can be said to be *trivially compatible* if and only if the operators that represent them commute with each other. More generally, this applies to any finite set of self-adjoint operators $\{\widehat{A}_1, \widehat{A}_2, \ldots, \widehat{A}_k\}$ that commute in pairs. However, there are many pairs of self-adjoint operators that do not commute, and hence many pairs of observables that are not trivially compatible. The question then arises if there are any pairs of observables that are of this type but which are still compatible in the sense that they can nevertheless be measured simultaneously in some way. Clearly this depends on what is meant by 'measured simultaneously'.

Von Neumann studied this question carefully and came to the conclusion that if observables A and B are simultaneously measurable, then necessarily their associated quantum operators commute, *i.e.*, $[\widehat{A}, \widehat{B}] = 0$. His only assumptions were (i) the general Rules of quantum theory as we have stated them; (ii) the function-preserving condition Eq. (5.14); and (iii) that the probabilistic predictions of the results of simultaneous measurements are consistent with those of the results of individual measurements. By this he meant the following. Suppose, for simplicity, that the eigenvalues of $\widehat{A}$ and $\widehat{B}$ are discrete, and let $\mathrm{Prob}(A = a_i, B = b_j; \vec{\psi})$ be the probability that, in a joint measurement of A and B, the results a_i and b_j will be obtained respectively for $i = 1, 2, \ldots, M_A$, $j = 1, 2, \ldots, M_B$. Then von Neumann's consistency conditions are

$$\begin{aligned}
\sum_{j=1}^{M_B} \mathrm{Prob}(A=a_i, B=b_j;\vec{\psi}) &= \mathrm{Prob}(A=a_i;\vec{\psi}) && (6.37)\\
\sum_{i=1}^{M_A} \mathrm{Prob}(A=a_i, B=b_j;\vec{\psi}) &= \mathrm{Prob}(B=b_j;\vec{\psi}) && (6.38)
\end{aligned}$$

i.e., the 'marginal' probabilities associated with the joint probability are equal to the single-quantity probabilities.

This line of argument leads to the strong statement (von Neumann 1971):

> "A necessary and sufficient condition for two observables A and B to be simultaneously measurable is that the operators that represent them should satisfy $[\widehat{A}, \widehat{B}] = 0$."

This means, for example, that it is impossible to make a simultaneous measurement of the position and the momentum of a particle. Such a view is often espoused but it has been strongly challenged, in particular by Park & Margenau (1968) who, as mentioned in Section 5.2, interpreted the term 'measurement' in such a way that a simultaneous measurement of non-commuting quantities *is* possible. They avoided the impact of the von Neumann theorem by showing that an alternative to imposing the condition $[\widehat{A}, \widehat{B}] = 0$ is to weaken Rule 2 of quantum theory to say that only *certain* observable quantities are represented by self-adjoint operators.

The whole concept of simultaneous measurement for non-commuting observables is clearly fraught with potential ambiguity, and it is safer to stick to the straightforward case of trivially-compatible observables. As we shall see in Section 7.3, and contrary to what is sometimes said, the uncertainty relations $\Delta x \Delta p \geq \hbar/2$ can be given a meaning within the instrumentalist approaches to quantum theory that does not depend at all on whether they can, or cannot, be subject to a simultaneous measurement.

6.4 Time Development

6.4.1 The Deterministic Evolution

The fourth quantum rule stated in Section 5.1 describes how a quantum state evolves in time. The first thing to notice about the Schrödinger equation Eq. (5.2) is that it is *linear*. Thus if $\vec{\psi}_{t_1}$ and $\vec{\phi}_{t_1}$ evolve into $\vec{\psi}_t$ and $\vec{\phi}_t$ respectively during time $t - t_1$, then $\alpha\vec{\psi}_{t_1} + \beta\vec{\phi}_{t_1}$ evolves into the state $\alpha\vec{\psi}_t + \beta\vec{\phi}_t$.

The second crucial property of the Schrödinger equation is that it is a *first-order* differential equation in time. Hence the state at any given time t_1 determines uniquely the state at any later (or earlier) time t. Thus quantum theory is a deterministic structure in the general sense described in Section 4.2.

To make this statement manifest we start by rewriting Eq. (5.2) in Dirac notation as

$$i\hbar\frac{d}{dt}\,|\psi_t\rangle = \widehat{H}\,|\psi_t\rangle \tag{6.39}$$

where $|\psi_t\rangle$ denotes the state at time t. The rigorous meaning of this differential equation is that, for any vector $|\phi\rangle \in \mathcal{H}$, the complex numbers $\langle\phi|\psi_t\rangle$ satisfy the ordinary differential equation

$$i\hbar\frac{d}{dt}\langle\phi|\psi_t\rangle = \langle\phi|\,\widehat{H}\,|\psi_t\rangle. \tag{6.40}$$

If we assume that the Hamiltonian operator $\widehat{H}$ has a discrete set of eigenvalues $\{E_1, E_2, \ldots, E_M\}$ with the associated spectral representation

$$\widehat{H} = \sum_{m=1}^{M} E_m \widehat{P}_m, \tag{6.41}$$

then Eq. (6.40) can be rewritten as

$$i\hbar\frac{d}{dt}\langle\phi|\psi_t\rangle = \sum_{m=1}^{M} E_m\,\langle\phi|\,\widehat{P}_m\,|\psi_t\rangle \tag{6.42}$$

where $\widehat{P}_m$ projects onto the eigenspace of $\widehat{H}$ with eigenvalue E_m.

Now, since Eq. (6.42) is true for all $|\phi\rangle$, it is true in particular for vectors of the form $\widehat{P}_n\,|\chi\rangle$, and hence, for all $\langle\chi|$,

$$i\hbar\frac{d}{dt}\langle\chi|\,\widehat{P}_n\,|\psi_t\rangle = \sum_{m=1}^{M} E_m\langle\chi|\,\widehat{P}_n\widehat{P}_m\,|\psi_t\rangle. \tag{6.43}$$

However, the spectral projectors $\widehat{P}_n$ satisfy the equation $\widehat{P}_n\widehat{P}_m = \delta_{mn}\widehat{P}_n$, and hence Eq. (6.43) becomes the collection of differential equations

$$i\hbar\frac{d}{dt}\langle\chi|\,\widehat{P}_n\,|\psi_t\rangle = E_n\langle\chi|\,\widehat{P}_n\,|\psi_t\rangle \tag{6.44}$$

for $n = 1, 2, \ldots, M$, and for all $|\chi\rangle \in \mathcal{H}$.

These first-order differential equations can be solved immediately to give the values of the matrix elements $\langle\chi|\,\widehat{P}_n\,|\psi_{t_2}\rangle$ at time t_2 in terms of their values at some other time t_1. The result is

$$\langle\chi|\,\widehat{P}_n\,|\psi_{t_2}\rangle = e^{-\frac{i}{\hbar}E_n(t_2-t_1)}\langle\chi|\,\widehat{P}_n\,|\psi_{t_1}\rangle \quad n = 1, 2, \ldots, M \tag{6.45}$$

which, when substituted into the expansion

$$\langle\chi|\psi_{t_2}\rangle = \sum_{m=1}^{M}\langle\chi|\,\widehat{P}_m\,|\psi_{t_2}\rangle \tag{6.46}$$

gives

$$\langle\chi|\psi_{t_2}\rangle = \sum_{m=1}^{M} e^{-\frac{i}{\hbar}E_m(t_2-t_1)}\langle\chi|\,\widehat{P}_m\,|\psi_{t_1}\rangle. \tag{6.47}$$

Since this is true for all $\langle\chi|$, we get the explicit expression

$$|\psi_{t_2}\rangle = \sum_{m=1}^{M} e^{-\frac{i}{\hbar}E_m(t_2-t_1)}\widehat{P}_m\,|\psi_{t_1}\rangle \tag{6.48}$$

for the state at time t_2 in terms of the state at time t_1.

Note that, using the definition Eq. (5.28) of a function of a self-adjoint operator applied to the spectral representation Eq. (6.41) of $\widehat{H}$, Eq. (6.48) can be written succinctly as

$$|\psi_{t_2}\rangle = e^{-\frac{i}{\hbar}\widehat{H}(t_2-t_1)}\,|\psi_{t_1}\rangle. \tag{6.49}$$

Thus the state at time t_2 can be found from the state $|\psi_{t_1}\rangle$ at any other time t_1 by acting on it with the linear operator

$$\widehat{U}(t_2, t_1) := e^{-\frac{i}{\hbar}\widehat{H}(t_2-t_1)}. \tag{6.50}$$

This is an example of a *unitary* operator, the general definition of which is given in Section 7.1.

These results can be generalised to include a time-dependent Hamiltonian, and when $\widehat{H}$ has a continuous spectrum. In all cases, Eq. (6.49) shows that the quantum-mechanical time evolution of a state follows the general pattern discussed in Section 4.2. In this sense, quantum theory is as deterministic as is classical physics. A far more problematic type of time evolution is the sudden change in state vector that is often postulated to accompany a measurement of an observable. This so-called *reduction* of the state vector is discussed in Section 8.3.

6.4.2 A Simple Example

Some of the ideas concerning time evolution can be usefully illustrated with a simple example of a model system using 2×2 matrices.

Problem

Suppose that a particular dynamical aspect of the electron is described by the Hamiltonian operator

$$\widehat{H} = \widehat{\mathbb{1}} + \alpha\sigma_y$$

where α is some real number, σ_y is the Pauli spin-matrix $\begin{pmatrix} 0 & -i \\ i & 0 \end{pmatrix}$, and $\widehat{\mathbb{1}}$ is the unit 2×2 matrix.

(i) If a measurement is made of this observable, what are the possible results that could be obtained?

(ii) If the state vector at some time $t = 0$ is $|\psi\rangle = \begin{pmatrix} 1 \\ 0 \end{pmatrix}$, what will it be at some later time t?

(iii) How do the probabilities of obtaining various measurement results for S_z depend on time?

Answer

(i) We have $\widehat{H} = \widehat{\mathbb{1}} + \alpha\sigma_y = \begin{pmatrix} 1 & -i\alpha \\ i\alpha & 1 \end{pmatrix}$ and, by the rules of quantum theory, the only possible result of a measurement of the observable is one of the eigenvalues of this operator. These are the solutions λ of the characteristic equation $\det\begin{pmatrix} 1-\lambda & -i\alpha \\ i\alpha & 1-\lambda \end{pmatrix} = 0$. Thus $(1-\lambda)^2 - \alpha^2 = 0$, and so $\lambda = 1 \pm \alpha$.

(ii) The time evolution of any state satisfies the time-dependent Schrödinger equation $i\hbar\frac{d|\psi_t\rangle}{dt} = \widehat{H}\,|\psi_t\rangle$ which, in the present case, reads

$$i\hbar\frac{d}{dt}\begin{pmatrix} a(t) \\ b(t) \end{pmatrix} = \begin{pmatrix} 1 & -i\alpha \\ i\alpha & 1 \end{pmatrix}\begin{pmatrix} a(t) \\ b(t) \end{pmatrix}$$

where $\begin{pmatrix} a(t) \\ b(t) \end{pmatrix}$ denotes the state at time t. Therefore

$$i\hbar\begin{pmatrix} \dot{a}(t) \\ \dot{b}(t) \end{pmatrix} = \begin{pmatrix} a(t) - i\alpha b(t) \\ i\alpha a(t) + b(t) \end{pmatrix} \text{ which implies } \begin{array}{l} i\hbar\dot{a}(t) = a(t) - i\alpha b(t) \\ i\hbar\dot{b}(t) = i\alpha a(t) + b(t) \end{array}$$

and so $i\hbar\frac{d}{dt}(a+ib) = (1-\alpha)(a+ib)$ and $i\hbar\frac{d}{dt}(a-ib) = (1+\alpha)(a-ib)$. These equations have the solution

$$a+ib = Ae^{-it(1-\alpha)/\hbar}, \quad \text{and} \quad a-ib = Be^{-it(1+\alpha)/\hbar}$$

where $A := (a+ib)|_{t=0}$ and $B := (a-ib)|_{t=0}$.

But we are given that $|\psi_{t=0}\rangle = \binom{1}{0}$, *i.e.*, $a(0)=1$ and $b(0)=0$. Hence $A = B = 1$ and $a(t) = e^{-it/\hbar}\cos\frac{\alpha t}{\hbar}$, $b(t) = e^{-it/\hbar}\sin\frac{\alpha t}{\hbar}$. Thus, after time t, the state $\binom{1}{0}$ has evolved into

$$|\psi_t\rangle = e^{-it/\hbar}\begin{pmatrix}\cos\frac{\alpha t}{\hbar}\\ \sin\frac{\alpha t}{\hbar}\end{pmatrix}.$$

Alternative Method

Expand $\binom{1}{0}$ in terms of the eigenvectors $\binom{1}{i}$ and $\binom{1}{-i}$ of $\widehat{H}$ corresponding to the eigenvalues $1+\alpha$ and $1-\alpha$ respectively: $\binom{1}{0} = \frac{1}{2}\binom{1}{i} + \frac{1}{2}\binom{1}{-i}$. Then $|\psi_t\rangle = e^{-it\widehat{H}/\hbar}\,|\psi_0\rangle$ which, in our case, means that

$$\begin{aligned}|\psi_0\rangle = \begin{pmatrix}1\\0\end{pmatrix} \mapsto |\psi_t\rangle &= \frac{1}{2}e^{-it\widehat{H}/\hbar}\begin{pmatrix}1\\i\end{pmatrix} + \frac{1}{2}e^{-it\widehat{H}/\hbar}\begin{pmatrix}1\\-i\end{pmatrix}\\ &= \frac{1}{2}e^{-it(1+\alpha)/\hbar}\begin{pmatrix}1\\i\end{pmatrix} + \frac{1}{2}e^{-it(1-\alpha)/\hbar}\begin{pmatrix}1\\-i\end{pmatrix}\end{aligned}$$

where we have used the definition of a function of a self-adjoint operator. Thus

$$|\psi_t\rangle = e^{-it/\hbar}\begin{pmatrix}\cos\frac{\alpha t}{\hbar}\\ \sin\frac{\alpha t}{\hbar}\end{pmatrix}$$

as before.

(iii) The probabilities for obtaining the different possible results are computed using the alternative to Rule 3 discussed above. Thus (writing the vectors $\binom{1}{0}$ and $\binom{0}{1}$ as $\langle\uparrow|$ and $\langle\downarrow|$ respectively) we have

$$\text{Prob}(S_z = \frac{\hbar}{2}; |\psi_t\rangle) = |\langle\uparrow|\psi_t\rangle|^2 = \left|\left\langle\begin{pmatrix}1\\0\end{pmatrix}, \begin{pmatrix}\cos\frac{\alpha t}{\hbar}\\ \sin\frac{\alpha t}{\hbar}\end{pmatrix}\right\rangle\right|^2 = \cos^2\frac{\alpha t}{\hbar}$$

and

$$\text{Prob}(S_z = -\frac{\hbar}{2}; |\psi_t\rangle) = |\langle\downarrow|\psi_t\rangle|^2 = \left|\left\langle\begin{pmatrix}1\\0\end{pmatrix}, \begin{pmatrix}\cos\frac{\alpha t}{\hbar}\\ \sin\frac{\alpha t}{\hbar}\end{pmatrix}\right\rangle\right|^2 = \sin^2\frac{\alpha t}{\hbar}.$$

Note that, as expected, these two probabilities sum to one.

6.4.3 Conserved Quantities

A good example of the application of Eq. (6.49) is the theory of conserved quantities. In classical physics, to say that a physical quantity $A(x,p)$ is conserved, means that its value does not change in time as the point $(x(t),p(t))$ moves around the state space. That is

$$0 = \frac{dA}{dt}(x(t),p(t)) \equiv \frac{\partial A(x,p)}{\partial x}\frac{dx(t)}{dt} + \frac{\partial A(x,p)}{\partial p}\frac{dp(t)}{dt}. \tag{6.51}$$

Such a definition cannot be used in quantum theory since it is usually not meaningful to speak of an observable 'having' a particular value. Rather, a typical state will be a linear superposition of eigenstates of the associated operator. However, if it happens that the state of the system at some initial time $t = 0$ *is* an eigenstate $|a\rangle$ of $\widehat{A}$, it is meaningful to ask whether the time-evolved form is also an eigenstate of $\widehat{A}$, and with the same eigenvalue. Thus $\widehat{A}\,|a\rangle = a\,|a\rangle$, and we are interested in

$$|a,t\rangle := e^{-\frac{i}{\hbar}\widehat{H}t}\,|a\rangle. \tag{6.52}$$

The obvious way of finding out if any given vector is an eigenvector of some operator, is to act on it with the operator and see what happens. In the present case, this gives the chain of calculations

$$\begin{aligned}
\widehat{A}\,|a,t\rangle &= \widehat{A}e^{-\frac{i}{\hbar}\widehat{H}t}\,|a\rangle = [\,\widehat{A}, e^{-\frac{i}{\hbar}\widehat{H}t}\,]\,|a\rangle + e^{-\frac{i}{\hbar}\widehat{H}t}\widehat{A}\,|a\rangle \\
&= [\,\widehat{A}, e^{-\frac{i}{\hbar}\widehat{H}t}\,]\,|a\rangle + ae^{-\frac{i}{\hbar}\widehat{H}t}\,|a\rangle \\
&= [\,\widehat{A}, e^{-\frac{i}{\hbar}\widehat{H}t}\,]\,|a\rangle + a\,|a,t\rangle.
\end{aligned} \tag{6.53}$$

This motivates saying that an observable A in a quantum theory is *conserved* if the corresponding operator commutes with the Hamiltonian; *i.e.*,

$$[\,\widehat{A}, \widehat{H}\,] = 0. \tag{6.54}$$

For $[\,\widehat{A}, \widehat{H}\,] = 0$ implies $[\,\widehat{A}, \mathcal{F}(\widehat{H})\,] = 0$ for any function $\mathcal{F}(\widehat{H})$ [Exercise] and, in particular, $[\,\widehat{A}, e^{-\frac{i}{\hbar}\widehat{H}t}\,] = 0$. Hence Eq. (6.53) gives

$$\widehat{A}\,|a,t\rangle = a\,|a,t\rangle, \tag{6.55}$$

so that $|a,t\rangle$ is an eigenvector of $\widehat{A}$ with the same eigenvalue a as the initial state $|a\rangle = |a,0\rangle$. In these circumstances, we say that the value a of A possessed by the system is *conserved*.

Note that if the eigenvalue a is non-degenerate then $|a, t\rangle$ must equal $\beta(t)\,|a\rangle$ for some complex phase factor $\beta(t)$. However, if the eigenvalue a is *degenerate* then all that is shown by the above theorem is that $|a, t\rangle$ lies somewhere in the eigenspace of eigenvectors with this eigenvalue.

6.4.4 The Time Development of a Mixed State

Finally, let us say a few words about how the time-dependent Schrödinger equation Eq. (6.39) can be extended to mixed states. Let us start by considering the special case in which the density matrix $\widehat{\rho}$ is just a single projection operator $|\psi\rangle\langle\psi|$. The time evolution of the ket $|\psi\rangle$ given by Eq. (6.49) implies that the appropriate evolution for the projection operator $\widehat{\rho}(t_1) := |\psi_{t_1}\rangle\langle\psi_{t_1}|$ is

$$\widehat{\rho}(t_1) \mapsto \widehat{\rho}(t) := e^{\frac{-i}{\hbar}\widehat{H}(t-t_1)}\,\widehat{\rho}(t_1)\,e^{\frac{i}{\hbar}\widehat{H}(t-t_1)}. \tag{6.56}$$

However, any density matrix can be written as an affine sum of projection operators of this type, and therefore, if we assume that the time evolution preserves this affine structure (and this *is* an assumption), the evolution of a general density matrix $\widehat{\rho}$ will be described by Eq. (6.56).

A useful result arises by differentiating both sides of Eq. (6.56) with respect to the time label t. Using the spectral theorem for $\widehat{H}$ it is easy to see that this gives the following differential equation for the time evolution of $\widehat{\rho}$:

$$i\hbar\frac{d\widehat{\rho}(t)}{dt} = [\,\widehat{H}, \widehat{\rho}(t)\,]. \tag{6.57}$$

This is a direct analogue and extension of the time-dependent Schrödinger equation Eq. (6.39) for pure states.

Chapter 7

UNITARY OPERATORS IN QUANTUM THEORY

7.1 Unitary Operators

After self-adjoint operators, the most important operators in quantum theory are those that are unitary; indeed, we have already seen one such in the guise of the time-evolution operator Eq. (6.50). Unitary operators appear naturally whenever there is any *symmetry* of the physical system, usually in the form of a group of transformations that leaves invariant various key structures in the theory. Such a group might be associated with the internal properties of the system (such as, for example, the $SU2$ or $SU3$ symmetries of strong-interaction hadronic physics), or it could come from special properties of the space-time in which the system resides. A good example of the latter is the Poincaré group that leaves invariant the metric line-element of the Minkowski space-time of special relativity.

Group operations of this type are of major importance in many applications of quantum theory, including elementary particle physics, relativistic quantum field theory, atomic and molecular spectra, and the structure of crystal lattices. As we shall see, ideas of symmetry also lead directly to a justification of the canonical commutation relation $[\widehat{x}, \widehat{p}] = i\hbar\,\widehat{\mathbb{1}}$ that determines the explicit form of the state-space and operators of elementary wave mechanics.

A unitary operator is one that preserves the lengths of vectors and the 'angles' between them. It is thus an analogue of the idea of a *rotation* in elementary vector calculus. More precisely:

Definition An operator $\widehat{U}$ on a Hilbert space $\mathcal{H}$ is *unitary* if (i) it is invertible; and (ii) it preserves all scalar products, *i.e.*,

$$\langle \widehat{U}\vec{\psi}, \widehat{U}\vec{\phi}\rangle = \langle \vec{\psi}, \vec{\phi}\rangle \tag{7.1}$$

for all $\vec{\psi}, \vec{\phi} \in \mathcal{H}$.

Comments

1. The invertibility requirement and Eq. (7.1) together are equivalent to the conditions [Exercise]

$$\widehat{U}\widehat{U}^\dagger = \widehat{U}^\dagger\widehat{U} = \widehat{\mathbb{1}}, \tag{7.2}$$

which can serve therefore as an alternative definition of a unitary operator.

2. The *norm* of a vector is preserved since $\| \widehat{U}\vec{\psi} \|^2 = \langle \widehat{U}\vec{\psi}, \widehat{U}\vec{\psi}\rangle = \langle \vec{\psi}, \vec{\psi}\rangle = \| \vec{\psi} \|^2$.

3. The operator conditions $\widehat{\mathbb{1}} = \widehat{U}\widehat{U}^\dagger$ and $\widehat{\mathbb{1}} = \widehat{U}^\dagger\widehat{U}$ can be converted into conditions on the matrix elements $U_{ij} := \langle e_i | \widehat{U} | e_j\rangle$ of $\widehat{U}$ defined with respect to an orthonormal basis $\{ |e_1\rangle, |e_2\rangle, \ldots, |e_N\rangle\}$ of $\mathcal{H}$. For example, taking matrix elements of the equation $\widehat{\mathbb{1}} = \widehat{U}\widehat{U}^\dagger$ gives

$$\langle e_i | e_j\rangle = \langle e_i | \widehat{U}\widehat{U}^\dagger | e_j\rangle = \sum_{k=1}^{N} \langle e_i | \widehat{U} | e_k\rangle\langle e_k | \widehat{U}^\dagger | e_j\rangle = \sum_{k=1}^{N} \langle e_i | \widehat{U} | e_k\rangle\langle e_j | \widehat{U} | e_k\rangle^* \tag{7.3}$$

so that

$$\delta_{ij} = \sum_{k=1}^{N} U_{ik}\, U_{jk}^*, \tag{7.4}$$

and a similar expression can be obtained using the equation $\widehat{\mathbb{1}} = \widehat{U}^\dagger\widehat{U}$. Thus the matrix that represents $\widehat{U}$ is a unitary matrix in the usual sense.

The key results in Section 3.3 for self-adjoint operators all have analogues in the unitary case. In particular:

Theorem

1. The eigenvalues of a unitary operator $\widehat{U}$ are complex numbers of modulus one.

2. Eigenvectors corresponding to different eigenvalues are orthogonal.

Proof

1. Let $\widehat{U}\vec{u} = a\vec{u}$. Then

$$\langle \vec{u}, \vec{u} \rangle = \langle \widehat{U}\vec{u}, \widehat{U}\vec{u} \rangle = \langle a\vec{u}, a\vec{u} \rangle = a^* a \langle \vec{u}, \vec{u} \rangle \tag{7.5}$$

which, since $\| \vec{u} \| \neq 0$, implies that $a^* a = 1$.

2. Let $\widehat{U}\vec{u_1} = a_1 \vec{u_1}$ and $\widehat{U}\vec{u_2} = a_2 \vec{u_2}$. Then

$$\langle \vec{u_2}, \vec{u_1} \rangle = \langle \widehat{U}\vec{u_2}, \widehat{U}\vec{u_1} \rangle = \langle a_2 \vec{u_2}, a_1 \vec{u_1} \rangle = a_2^* a_1 \langle \vec{u_2}, \vec{u_1} \rangle = \frac{a_1}{a_2} \langle \vec{u_2}, \vec{u_1} \rangle. \tag{7.6}$$

However, $a_1 \neq a_2$, and hence Eq. (7.6) gives $\langle \vec{u_2}, \vec{u_1} \rangle = 0$, as claimed. **QED**

Comments

1. The crucial completeness theorem also applies to a unitary operator $\widehat{U}$. Thus its eigenvectors $|e^{i\theta_m}, j\rangle$ (where j is a degeneracy label) form a basis set for the Hilbert space, and any vector can be expanded as (*cf.* Eq. (3.53))

$$|\psi\rangle = \sum_{m=1}^{M} \sum_{j=1}^{d(m)} \langle e^{i\theta_m}, j | \psi \rangle \, |e^{i\theta_m}, j\rangle \tag{7.7}$$

where $d(m)$ is the degeneracy of the mth eigenvalue $e^{i\theta_m}$. The associated resolution of the identity is (*cf.* Eq. (3.54))

$$\widehat{\mathbb{1}} = \sum_{m=1}^{M} \sum_{j=1}^{d(m)} |e^{i\theta_m}, j\rangle\langle e^{i\theta_m}, j| \,, \tag{7.8}$$

and the spectral representation of $\widehat{U}$ is (*cf.* Eq. (3.55) and Eq. (3.60))

$$\widehat{U} = \sum_{m=1}^{M} \sum_{j=1}^{d(m)} e^{i\theta_m} \, |e^{i\theta_m}, j\rangle\langle e^{i\theta_m}, j| = \sum_{m=1}^{M} e^{i\theta_m} \, \widehat{P}_m \tag{7.9}$$

where

$$\widehat{P}_m := \sum_{j=1}^{d(m)} |e^{i\theta_m}, j\rangle\langle e^{i\theta_m}, j| \tag{7.10}$$

is the projection operator onto the subspace of vectors with eigenvalue $e^{i\theta_m}$.

2. Like self-adjoint operators, unitary operators can have continuous eigenvalues as well as discrete ones. Operators of this type can be handled in ways that are very similar to those discussed in section 6.2. In particular, there is an appropriate generalisation of the spectral theorem to give a spectral representation of the form (*cf.* Eq. (6.15))

$$\langle\vec{\psi}, \widehat{U}\vec{\phi}\rangle = \int_0^{2\pi} e^{i\theta}\, d_\theta\langle\vec{\psi}, \widehat{E}_\theta\vec{\phi}\rangle. \tag{7.11}$$

3. There is a deep connection between self-adjoint operators and unitary operators that is of considerable importance in quantum theory. The eigenvalues of a self-adjoint or a unitary operator are respectively a real number or a complex number of modulus one. On the other hand, if r is any real number then $|e^{ir}| = 1$. This motivates the following significant theorem:

Theorem

If $\widehat{A}$ is self-adjoint then $e^{i\widehat{A}}$ is a unitary operator.

Proof

For simplicity we shall consider only the case where the eigenvalue spectrum of the operator $\widehat{A}$ is discrete. Then $\widehat{A}$ has the spectral representation $\widehat{A}=\sum_{m=1}^{M} a_m\widehat{P}_m$, and hence, according to the definition Eq. (5.28) of a function of a self-adjoint operator, we have

$$e^{i\widehat{A}} = \sum_{m=1}^{M} e^{ia_m}\widehat{P}_m. \tag{7.12}$$

In turn, using the result $\widehat{P}_m^\dagger = \widehat{P}_m$, this implies that the adjoint $(e^{i\widehat{A}})^\dagger$ has the representation

$$\left(e^{i\widehat{A}}\right)^\dagger = \sum_{m=1}^{M} e^{-ia_m}\widehat{P}_m. \tag{7.13}$$

Then

$$
\begin{aligned}
e^{i\widehat{A}}\left(e^{i\widehat{A}}\right)^{\dagger} &= \sum_{m=1}^{M}\sum_{k=1}^{M} e^{i(a_m-a_k)}\widehat{P}_m\widehat{P}_k \\
&= \sum_{m=1}^{M}\sum_{k=1}^{M} e^{i(a_m-a_k)}\delta_{mk}\widehat{P}_k = \sum_{m=1}^{M}\widehat{P}_m = \widehat{\mathbb{1}}
\end{aligned}
$$

where we have used $\widehat{P}_m\widehat{P}_k = \delta_{mk}\widehat{P}_k$.

Similarly, it can be shown that $(e^{i\widehat{A}})^{\dagger}e^{i\widehat{A}}=\widehat{\mathbb{1}}$. **QED**

Comments

1. A very important role is played in quantum theory by one-parameter families of unitary operators $\widehat{U}(r)$, $r \in \mathbb{R}$, defined by

$$\widehat{U}(r) := e^{ir\widehat{A}} = \sum_{m=1}^{M} e^{ira_m}\widehat{P}_m \tag{7.14}$$

for some fixed self-adjoint operator $\widehat{A}$. It is easy to show from the spectral theorem that this family $r \mapsto \widehat{U}(r)$ of unitary operators satisfies the three conditions

$$\widehat{U}(0) = \widehat{\mathbb{1}} \tag{7.15}$$

$$\widehat{U}(r_1)\widehat{U}(r_2) = \widehat{U}(r_1 + r_2) \tag{7.16}$$

$$\widehat{U}(-r) = \widehat{U}(r)^{-1} \tag{7.17}$$

for all $r, r_1, r_2 \in \mathbb{R}$. In group theory language, these equations say that $r \mapsto \widehat{U}(r)$ is a unitary representation of the additive group of the real numbers. In particular, the time-evolution operator in Eq. (6.50) falls into this category if the initial time t_1 is chosen to be $t_1 = 0$.

2. As far as quantum theory is concerned the really important theorem (due to the mathematician Stone) is the converse of the above:

Theorem

Let $\{\widehat{U}(r), r \in \mathbb{R}\}$ be a collection of unitary operators labelled by a real number r. Suppose that this one-parameter family of operators satisfies the following three conditions:

1. The matrix element $\langle\phi|\,\widehat{U}(r)\,|\psi\rangle$ is a continuous function of r for all vectors $|\phi\rangle$, $|\psi\rangle \in \mathcal{H}$.

2. $\widehat{U}(0) = \widehat{\mathbb{1}}$

3. For all $r_1, r_2 \in \mathbb{R}$, $\widehat{U}(r_1)\widehat{U}(r_2) = \widehat{U}(r_1 + r_2)$.

Then there exists a unique self-adjoint operator $\widehat{A}$ such that

$$\widehat{U}(r) = e^{ir\widehat{A}} \tag{7.18}$$

and

$$i\widehat{A}\,|\psi\rangle = \lim_{r\to 0} \frac{\widehat{U}(r) - \widehat{\mathbb{1}}}{r}\,|\psi\rangle \quad \text{for all } |\psi\rangle \in \mathcal{H} \tag{7.19}$$

where the right hand side of Eq. (7.19) means strong convergence of vectors in the sense discussed briefly in Section 2.5.

Stone's theorem is too difficult to prove here (for example, see Reed & Simon (1972)) but its significance will be shown shortly. In practice, it is one of the most important ways whereby specific self-adjoint operators arise in quantum theory.

7.2 Some Applications in Quantum Theory

7.2.1 The Basic Role of Unitary Operators

Most unitary operators are not self-adjoint and therefore cannot represent physical observables directly. Their use in quantum theory is rather to quantify the arbitrariness in the association of vectors and self-adjoint operators with states and observables: a property that is far more important than might appear at a first glance.

Let $\widehat{U}$ be a unitary operator on the Hilbert space $\mathcal{H}$, and let $\widehat{A}$ be a self-adjoint operator. Then $\widehat{U}\widehat{A}\widehat{U}^{-1}$ $(=\widehat{U}\widehat{A}\widehat{U}^{\dagger})$ is also self-adjoint since $(\widehat{U}\widehat{A}\widehat{U}^{\dagger})^{\dagger} = (\widehat{U}^{\dagger})^{\dagger}\widehat{A}^{\dagger}\widehat{U}^{\dagger} = \widehat{U}\widehat{A}\widehat{U}^{\dagger}$.

Furthermore, if a is an eigenvalue of $\widehat{A}$ with eigenvector $\vec{u}$, then

$$\widehat{U}\widehat{A}\widehat{U}^{-1}(\widehat{U}\vec{u}) = \widehat{U}(\widehat{A}\vec{u}) = \widehat{U}(a\vec{u}) = a(\widehat{U}\vec{u}) \tag{7.20}$$

which shows that any eigenvalue of $\widehat{A}$ is also an eigenvalue of $\widehat{U}\widehat{A}\widehat{U}^{-1}$ for any unitary operator $\widehat{U}$. Conversely, using the same argument but with $\widehat{A}$ replaced with the self-adjoint operator $(\widehat{U}\widehat{A}\widehat{U}^{-1})$, and $\widehat{U}$ replaced with the unitary operator $\hat{U}^{-1}$, we see that any eigenvalue of $\hat{U}\hat{A}\hat{U}^{-1}$ is necessarily an eigenvalue of $\hat{U}^{-1}(\hat{U}\hat{A}\hat{U}^{-1})\hat{U} \equiv \hat{A}$. Thus $\hat{U}\hat{A}\hat{U}^{-1}$ has exactly the same eigenvalues as the original operator $\hat{A}$.

Finally, since $\hat{U}$ is unitary, we have, for all $\vec{\psi}, \vec{\phi} \in \mathcal{H}$,

$$|\langle \hat{U}\vec{\psi}, \hat{U}\vec{\phi}\rangle| = |\langle \vec{\psi}, \vec{\phi}\rangle| \tag{7.21}$$

and,

$$\langle \hat{U}\vec{\psi}, (\hat{U}\hat{A}\hat{U}^{-1})\,\hat{U}\vec{\phi}\rangle = \langle \hat{U}\vec{\psi}, \hat{U}\hat{A}\vec{\phi}\rangle = \langle \vec{\psi}, \hat{A}\vec{\phi}\rangle. \tag{7.22}$$

Hence, using Rule 3 in Section 5.1 and equation Eq. (5.29), we obtain the result that if a specific association has been made of

$$\begin{aligned} \text{Physical state} \quad &\longmapsto \quad \text{state vector } \vec{\psi} \in \mathcal{H} \\ \text{Physical observable } A \quad &\longmapsto \quad \text{self-adjoint operator } \hat{A} \text{ acting on } \mathcal{H} \end{aligned}$$

then the same physical predictions will be obtained if the following association is used instead

$$\text{Physical state} \quad \longmapsto \quad \text{state vector } \hat{U}\vec{\psi} \in \mathcal{H} \tag{7.23}$$

$$\text{Physical observable } A \quad \longmapsto \quad \text{self-adjoint operator } \hat{U}\hat{A}\hat{U}^{-1} \text{ acting on } \mathcal{H} \tag{7.24}$$

for any unitary operator $\hat{U}$. Thus we arrive at the important conclusion that the association of physical quantities with their mathematical representatives is defined only up to arbitrary transformations of the type above.

More generally, we could consider arbitrary maps $\vec{\psi} \mapsto \vec{\psi}'$ with the property that, for all vectors $\vec{\psi}, \vec{\phi} \in \mathcal{H}$,

$$|\langle \vec{\psi}', \vec{\phi}'\rangle| = |\langle \vec{\psi}, \vec{\phi}\rangle|. \tag{7.25}$$

Such maps are clearly of interest in quantum physics since they leave invariant probabilistic predictions like Eq. (5.29). However, there seems to be no *prima facie* reason why a map from $\mathcal{H}$ to $\mathcal{H}$ satisfying Eq. (7.25) should even be *linear*, let alone be derived from the action of a unitary operator. This makes the following theorem all the more remarkable.

Theorem (Wigner)

Let $\vec{\psi} \mapsto \vec{\psi}'$ be any map from a Hilbert space $\mathcal{H}$ to itself that is (i) invertible; and (ii) such as to satisfy $|\langle \vec{\psi}, \vec{\phi}\rangle| = |\langle \vec{\psi}', \vec{\phi}'\rangle|$ for all $\vec{\psi}, \vec{\phi} \in \mathcal{H}$.

Then there exists an operator $\widehat{U} : \mathcal{H} \to \mathcal{H}$ (defined uniquely up to an arbitrary phase factor) that is either unitary or anti-unitary and is such that $\vec{\psi}' = \widehat{U}\vec{\psi}$ for all vectors $\vec{\psi} \in \mathcal{H}$.

Comments

1. An *anti-unitary* operator $\widehat{A}$ is defined to be a map from $\mathcal{H}$ to itself that is

i) anti-linear in the sense that (*cf.* Eq. (2.28))

$$\widehat{A}(\alpha\vec{\psi} + \beta\vec{\phi}) = \alpha^* \widehat{A}\vec{\psi} + \beta^* \widehat{A}\vec{\phi} \tag{7.26}$$

for all complex numbers α, β and vectors $\vec{\psi}, \vec{\phi} \in \mathcal{H}$; and

ii) such that for all vectors $\vec{\psi}, \vec{\phi} \in \mathcal{H}$, (*cf.* Eq. (7.1))

$$\langle \widehat{A}\vec{\psi}, \widehat{A}\vec{\phi} \rangle = \langle \vec{\phi}, \vec{\psi} \rangle \quad (= \langle \vec{\psi}, \vec{\phi} \rangle^*). \tag{7.27}$$

2. The square of an anti-unitary operator $\widehat{A}$ is unitary since $\langle \widehat{A}^{\,2}\vec{\psi}, \widehat{A}^{\,2}\vec{\phi} \rangle = \langle \widehat{A}\vec{\phi}, \widehat{A}\vec{\psi} \rangle = \langle \vec{\psi}, \vec{\phi} \rangle$. Similarly, if $\widehat{U}$ is unitary so is $\widehat{U}^{\,2}$.

3. A well-known example of an anti-unitary operator is the time-reversal operator $\widehat{T}$ in wave mechanics which is defined by $(\widehat{T}\psi)(x) := \psi(x)^*$.

7.2.2 Displaced Observers and the Canonical Commutation Relations

Situations often arise in quantum theory in which, for clear physical reasons, we expect to get different (but equivalent in the sense of Eq. (7.24)) mathematical representations of the same underlying physical states and observables. We shall now demonstrate the application of these ideas in the context of displaced observers. To facilitate the presentation the material is broken up into a series of 11 steps.

1. Consider an isolated quantum system that is under observation by two observers O_1 and O_2, and with O_2 displaced a distance a along the x-axis from where O_1 is standing (see Figure 7.1) at the origin of coordinates. Assume that both use the same quantum-mechanical rules for

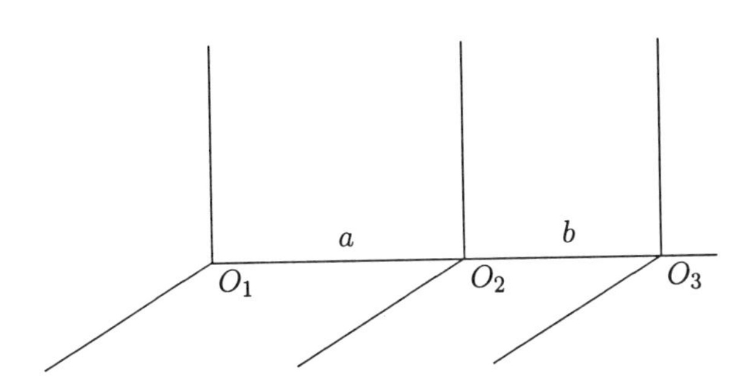

Figure 7.1: A displaced set of observers

associating physical states with vectors in a Hilbert space $\mathcal{H}$. Then it can be anticipated that a given physical state of the system will be described by them using different state vectors (but in the same Hilbert space) since the system looks different in certain ways when viewed from their different perspectives.

However, displacing the first observer a to reach O_2 should be the same as displacing the system a distance $-a$ towards O_1, and hence, because of the *homogeneous* nature of physical space, the quantum-mechanical results of measurements (of the *same* observable) made on the system should be the same for the two observers. In particular, if the vectors assigned by O_1 and O_2 are written as $|\psi\rangle$ and $|\psi\rangle_a$ respectively, we must have

$$|\langle\psi|\phi\rangle|^2 = |_a\langle\psi|\phi\rangle_a|^2 \tag{7.28}$$

in order that the probabilistic predictions be unchanged.

2. It then follows from Wigner's theorem that there exists a unitary or anti-unitary operator $\widehat{D}(a)$ such that

$$|\psi\rangle_a = \widehat{D}(a)\,|\psi\rangle. \tag{7.29}$$

3. Suppose now a third observer O_3 joins the experiment and stands a distance b from O_2 (see Figure 7.1). On physical grounds, the state $|\psi\rangle_{b+a} = \widehat{D}(b+a)\,|\psi\rangle$ assigned by O_3 should be the same[1] whether we

[1] Actually, this could be true only up to a phase factor. But it turns out that this can be absorbed in the definition of $\widehat{D}$.

start with O_1's state $|\psi\rangle$ and then go directly to where O_3 is standing at $x = b + a$, or whether we go first to $x = a$, pick up the state $|\psi\rangle_a$ that O_2 assigns, and then go from there to $x = b + a$. Thus

$$\begin{aligned} |\psi\rangle_{b+a} &= \widehat{D}(b+a)\,|\psi\rangle \\ &= \widehat{D}(b)\,|\psi\rangle_a = \widehat{D}(b)\widehat{D}(a)\,|\psi\rangle \end{aligned} \tag{7.30}$$

for all states $|\psi\rangle$. This means that, for all $a, b \in \mathbb{R}$,

$$\widehat{D}(b)\widehat{D}(a) = \widehat{D}(b+a). \tag{7.31}$$

4. In particular, inserting the special values $a = b = r/2$ gives $\widehat{D}(r) = \widehat{D}(r/2)\widehat{D}(r/2)$. But, by Comment 2 to Wigner's theorem, the product $\widehat{D}(r/2)\widehat{D}(r/2)$ is unitary irrespective of whether $\widehat{D}(r/2)$ is unitary or anti-unitary. Thus, for any $r \in \mathbb{R}$, $\widehat{D}(r)$ is in fact a unitary operator. Note how, at this point, we have sneaked in a certain assumption about the nature of physical space, *i.e.*, that any r can be written as $r = r/2 + r/2$.

5. Clearly $\widehat{D}(0) = \widehat{\mathbb{1}}$.

6. On physical grounds, we expect $\langle\phi|\psi\rangle_a = \langle\phi|\,\widehat{D}(a)\,|\psi\rangle$ to change continuously as the observer at a is displaced along the x-axis in a continuous fashion. Again, this is a reflection of our views on nature of physical space, *i.e.*, that it is a *continuum.*

7. We have seen that all three conditions for Stone's theorem are satisfied, and hence there exists a unique self-adjoint operator, denoted $\widehat{d}_x$ (and called the *generator* of displacements along the x-direction), such that

$$\widehat{D}(a) = e^{ia\widehat{d}_x}, \tag{7.32}$$

and where

$$i\widehat{d}_x\,|\psi\rangle = \lim_{a\to 0} \frac{\widehat{D}(a) - \widehat{\mathbb{1}}}{a}\,|\psi\rangle \tag{7.33}$$

for all[2] vectors $|\psi\rangle$.

[2]Strictly speaking, this equation only holds on the subset of vectors on which the unbounded operator $\widehat{d}_x$ is actually defined.

8. Under the transformation that takes the state $|\psi\rangle$ to $\widehat{D}(a)\,|\psi\rangle$, the corresponding assignment of self-adjoint operators to observables transforms as

$$\widehat{B} \mapsto \widehat{B}_a = \widehat{D}(a)\widehat{B}\widehat{D}(a)^{-1} \tag{7.34}$$

where $\widehat{B}$ and $\widehat{B}_a$ are the operators assigned by O_1 and O_2 respectively. Now if a is small we can expand the exponential in Eq. (7.32) and write[3] $\widehat{D}(a) = 1 + ia\widehat{d}_x + O(a^2)$. Then Eq. (7.34) becomes

$$\begin{aligned} \widehat{B}_a &= (1 + ia\widehat{d}_x)\widehat{B}(1 - ia\widehat{d}_x) + O(a^2) \\ &= \widehat{B} + ia[\,\widehat{d}_x, \widehat{B}\,] + O(a^2). \end{aligned} \tag{7.35}$$

Note the occurrence on the right hand side of a *commutator* of two operators. In fact, most important commutators in quantum theory have their origin in expansions like that in Eq. (7.35).

9. Consider in particular the x-component of the center of mass of the system as measured by O_1 from the origin of his reference frame where he is standing. Then, by Eq. (7.35), the operator assigned to this observable by the translated observer O_2 is

$$\widehat{x}_a = \widehat{x} + ia[\,\widehat{d}_x, \widehat{x}\,] + O(a^2). \tag{7.36}$$

On the other hand, the *same* physical observable will also be represented by O_2 by the operator $\widehat{x} + a$, since she uses $\widehat{x}$ to denote the position from the origin (the point $x = a$) of *her* reference frame. Thus $\widehat{x}_a = \widehat{x} + a$, and hence from Eq. (7.36), in the limit $a \to 0$,

$$-i = [\,\widehat{d}_x, \widehat{x}\,]. \tag{7.37}$$

10. The final step is to identify the operator $\widehat{d}_x$ with $\widehat{p}_x/\hbar$, where $\widehat{p}_x$ is the momentum along the x direction. This can be done by appealing to the classical limit of the theory, or by requiring consistency with the

[3] If $\mathcal{H}$ is finite-dimensional, this type of result is rigorous in the sense that one can write $e^{ia\widehat{d}_x} = \sum_{n=0}^{\infty}(ia\widehat{d}_x)^n/n!$, where the partial sums converge in an appropriate sense. Greater care is needed in the infinite-dimensional case but the basic idea is more or less correct. It can be made rigorous using Stone's theorem.

results of elementary wave mechanics. Thus we get the result that the states assigned by O_2 and O_1 are related by

$$|\psi\rangle_a = e^{ia\widehat{p}_x/\hbar}\,|\psi\rangle. \tag{7.38}$$

Comments

1. The argument above can be generalised to include translations along all three axes. If the associated self-adjoint generators are denoted $\widehat{d}_k$, $k = 1, 2, 3$, then Eq. (7.37) becomes

$$[\,\widehat{x}_j, \widehat{d}_k\,] = i\delta_{jk} \tag{7.39}$$

where $\widehat{x}_j$, $j = 1, 2, 3$ are the position operators associated with the three coordinate axes of the observer at the origin of coordinates. The same type of analysis applied to the measurement of momentum gives

$$[\,\widehat{p}_j, \widehat{p}_k\,] = 0, \tag{7.40}$$

and consideration of reference frames moving with different constant velocities with respect to each other gives

$$[\,\widehat{x}_j, \widehat{x}_k\,] = 0. \tag{7.41}$$

Thus we see that the famous canonical commutation relations of non-relativistic wave mechanics can actually be *derived* from the general quantum-mechanical formalism assuming only that the structure of physical space is homogeneous.

2. The entire analysis can be repeated for observers whose reference frames are rotated by a fixed amount with respect to each other. Assuming that physical space is *isotropic* (*i.e.*, there are no preferred directions) this leads to the existence of a set of 'rotation generators' $\widehat{R}_x$, $\widehat{R}_y$, $\widehat{R}_z$ that satisfy the commutation relations

$$[\,\widehat{R}_x, \widehat{R}_y\,] = i\widehat{R}_z \tag{7.42}$$

plus cyclic permutations. Thus, identifying $\widehat{R}_x$ with $\widehat{J}_x/\hbar$, we recover the familiar algebra satisfied by the angular momentum operators.

3. Let $\widehat{A}$ and $\widehat{B}$ be any self-adjoint operators that satisfy the commutation relation

$$[\widehat{A}, \widehat{B}] = i\hbar, \tag{7.43}$$

and let $|a\rangle$ be an eigenvector of $\widehat{A}$ with eigenvalue a. Then taking the diagonal matrix element of the left hand side of Eq. (7.43) with respect to $|a\rangle$ gives

$$\langle a|\,[\widehat{A}, \widehat{B}]\,|a\rangle = \langle a|\,\widehat{A}\widehat{B} - \widehat{B}\widehat{A}\,|a\rangle = (a-a)\langle a|\,\widehat{B}\,|a\rangle = 0, \tag{7.44}$$

whereas the matrix element of the right hand side of Eq. (7.43) is $i\hbar$. Dividing both sides by $i\hbar$ therefore proves the famous result $0 = 1$!

This contradiction seems to imply that no operators can exist that satisfy commutation relations Eq. (7.43). This would mean that at least one step leading to Eq. (7.37) is false; for example, perhaps the assumption that space is continuous (used in step 6) is incorrect. However, we know that operators satisfying Eq. (7.43) *do* in fact exist, namely the $\widehat{x}$ and $\widehat{p}$ operators of elementary wave mechanics. The answer to this apparent paradox is that the eigenvalue spectrum of both operators must be *continuous*, in which case the generalised eigenvectors are not normalisable, and then the scalar products in Eq. (7.44) have no meaning. This is a salutary example of how careful one needs to be when handling operators whose spectra are continuous!

Note that operators with a continuous spectrum can arise only if the Hilbert space $\mathcal{H}$ has an infinite dimension. Thus we see that *any* quantum theory that takes into account the position of a system in space (and which therefore leads to the canonical commutation relation Eq. (7.37)) must necessarily use an infinite-dimensional Hilbert space.

On the other hand there is no problem with the angular momentum commutation relations $[\widehat{J}_x, \widehat{J}_y] = i\hbar\widehat{J}_z$ since, if $|k\rangle$ is an eigenvector of $\widehat{J}_x$, the matrix-element equation becomes $0 = \langle k|\,\widehat{J}_z\,|k\rangle$, and there is no reason why a normalisable state should not satisfy this condition.

7.3 The Uncertainty Relations

7.3.1 Some Ideas From Elementary Statistics

Having derived the canonical commutation relations from the quantum-mechanical formalism we shall now consider some of the implications for the interpretation of the theory. This has been the subject of many conflicting statements over the years, and it is necessary to be precise about what the uncertainty relations actually refer to.

An expression like 'Δx' can have a variety of meanings that are frequently confused[4] with each other. The safest approach is to decide first on the general interpretative framework within which one is going to work, and then to consider the most appropriate definition for the 'uncertainty' of an observable in that scheme. In particular, in the pragmatic approach to quantum theory, the uncertainty ΔA in an observable A refers to a precise numerical measure of the spread of results when repeated measurements of A are made. We shall see that the uncertainty relations for these quantities can be *derived* from the formalism itself: they are not an extra ingredient that needs to be postulated over and above the rules given already.

The starting point is certain definitions that arise in the theory of elementary statistics. Thus, if a measurement of the observable A is predicted to yield results $a_1, a_2, \ldots, a_M$ with probabilities $\mathrm{p}_1, \mathrm{p}_2, \ldots, \mathrm{p}_M$, the *expected result* of A is defined as

$$\langle A \rangle := \sum_{i=1}^{M} a_i \, \mathrm{p}_i \tag{7.45}$$

and the *variance* of A is

$$\sigma^2 := \sum_{i=1}^{M} (a_i - \langle A \rangle)^2 \, \mathrm{p}_i \tag{7.46}$$

which can be rewritten as

$$\sigma^2 = \sum_{i=1}^{M} (a_i^2 + \langle A \rangle^2 - 2a_i \langle A \rangle) \, \mathrm{p}_i = \sum_{i=1}^{M} a_i^2 \, \mathrm{p}_i - \langle A \rangle^2. \tag{7.47}$$

---

[4]For an excellent analysis of this point see the discussion in Scheibe (1973).

Thus

$$\sigma^2 = \langle A^2 \rangle - \langle A \rangle^2 \tag{7.48}$$

is a measure of the predicted spread of results around the expected value $\langle A \rangle$ of A. The quantity σ is called the *standard deviation* of A.

In the approaches to quantum theory based on the relative-frequency interpretation of probability, this motivates the definition of the *uncertainty*, or *dispersion*, $\Delta_\psi A$ in the results of measuring the observable A as

$$(\Delta_\psi A)^2 := \langle (A - \langle A \rangle_\psi)^2 \rangle_\psi = \langle A^2 \rangle_\psi - \langle A \rangle_\psi^2 \tag{7.49}$$

where $\langle A \rangle_\psi = \langle \psi | \widehat{A} | \psi \rangle$. Thus

$$\Delta_\psi A = (\langle \psi | \widehat{A}^2 | \psi \rangle - \langle \psi | \widehat{A} | \psi \rangle^2)^{\frac{1}{2}} \tag{7.50}$$

where it is understood that the square root is always taken to be positive.

Comments

1. The uncertainty $\Delta_\psi A$ depends on the *state* $|\psi\rangle$ of the system. This point tends to get lost in some of the more heuristic discussion of the uncertainty relations.

2. If $|\psi\rangle$ is an eigenvector of $\widehat{A}$ then $\Delta_\psi A = 0$. This is consistent with our earlier result that if the state of a system is an eigenvector then, with probability one, the result of a measurement of A will be the corresponding eigenvalue. That is, the spread of the results about the expected result is predicted to be zero.

Conversely, using the Schwarz inequality below it can be shown [Exercise] that $\Delta_\psi A = 0$ implies that $|\psi\rangle$ is an eigenstate of $\widehat{A}$. Thus the uncertainty $\Delta_\psi A$ is a good measure of the extent to which a value of A is *not* 'possessed' by an individual system. This idea is developed further below.

7.3.2 The Schwarz Inequality

The uncertainty relations themselves can be derived from the following famous theorem in Hilbert space theory. For reasons of typographical elegance this is written using non-Dirac notation.

Theorem (Schwarz Inequality)

Any pair of vectors $\vec{\psi}, \vec{\phi}$ in a Hilbert space $\mathcal{H}$ satisfies the inequality

$$|\langle\vec{\psi}, \vec{\phi}\rangle| \leq \langle\vec{\psi}, \vec{\psi}\rangle^{\frac{1}{2}} \langle\vec{\phi}, \vec{\phi}\rangle^{\frac{1}{2}} \tag{7.51}$$

with equality holding if, and only if, the two vectors are linearly dependent.

Proof

The theorem is trivially true if $\vec{\psi} = 0$ or $\vec{\phi} = 0$. If both vectors are non-zero then, for any complex number α, the basic condition Eq. (2.52) in the definition of the scalar product shows that

$$\begin{aligned} 0 &\leq \langle\vec{\psi} + \alpha\vec{\phi}, \vec{\psi} + \alpha\vec{\phi}\rangle \\ &= \langle\vec{\psi}, \vec{\psi}\rangle + \alpha^*\alpha\langle\vec{\phi}, \vec{\phi}\rangle + \alpha^*\langle\vec{\phi}, \vec{\psi}\rangle + \alpha\langle\vec{\psi}, \vec{\phi}\rangle. \end{aligned} \tag{7.52}$$

In particular, this is true for the clever choice

$$\alpha = -\langle\vec{\phi}, \vec{\psi}\rangle / \langle\vec{\phi}, \vec{\phi}\rangle$$

which, when substituted into the inequality Eq. (7.52), gives

$$0 \leq \langle\vec{\psi}, \vec{\psi}\rangle + \frac{|\langle\vec{\phi}, \vec{\psi}\rangle|^2}{\langle\vec{\phi}, \vec{\phi}\rangle} - \frac{\langle\vec{\phi}, \vec{\psi}\rangle^*\langle\vec{\phi}, \vec{\psi}\rangle}{\langle\vec{\phi}, \vec{\phi}\rangle} - \frac{\langle\vec{\phi}, \vec{\psi}\rangle\langle\vec{\psi}, \vec{\phi}\rangle}{\langle\vec{\phi}, \vec{\phi}\rangle}.$$

The right hand side of this inequality is

$$\langle\vec{\psi}, \vec{\psi}\rangle + \frac{|\langle\vec{\phi}, \vec{\psi}\rangle|^2}{\langle\vec{\phi}, \vec{\phi}\rangle} - 2\frac{|\langle\vec{\phi}, \vec{\psi}\rangle|^2}{\langle\vec{\phi}, \vec{\phi}\rangle}$$

and so

$$0 \leq \langle\vec{\psi}, \vec{\psi}\rangle - \frac{|\langle\vec{\phi}, \vec{\psi}\rangle|^2}{\langle\vec{\phi}, \vec{\phi}\rangle}$$

so that

$$|\langle\vec{\phi}, \vec{\psi}\rangle|^2 \leq \langle\vec{\psi}, \vec{\psi}\rangle\langle\vec{\phi}, \vec{\phi}\rangle,$$

which proves the first part of the result.

If $\vec{\psi}$ and $\vec{\phi}$ are linearly dependent then $\vec{\psi} = \beta\vec{\phi}$ for some $\beta \in \mathbb{C}$, and substitution of this into Eq. (7.51) shows at once that the equality

holds. Conversely, if the equality holds in the final expression obtained in the argument above, then it must hold at every stage. In particular, it must hold in Eq. (7.52), and hence by the basic condition on the scalar product we must have $\vec{\psi}+\alpha\vec{\phi}=0$, which means that $\vec{\psi}$ and $\vec{\phi}$ are linearly dependent.
QED

Comments

1. In ordinary vector calculus the Schwarz inequality is simply the expression $|\mathbf{u}\cdot\mathbf{v}| \leq (\mathbf{u}\cdot\mathbf{u})^{\frac{1}{2}}\,(\mathbf{v}\cdot\mathbf{v})^{\frac{1}{2}}$ which, if $\cos\theta$ is the angle between $\mathbf{u}$ and $\mathbf{v}$, is simply the assertion that $|\cos\theta| \leq 1$.

2. The Schwarz inequality can be used to show [Exercise] that the norm defined by $\|\vec{\psi}\| := \langle\vec{\psi},\vec{\psi}\rangle^{\frac{1}{2}}$ satisfies the *triangle inequality*

$$\|\vec{\psi}+\vec{\phi}\| \leq \|\vec{\psi}\| + \|\vec{\phi}\| \tag{7.53}$$

for all vectors $\vec{\psi},\vec{\phi} \in \mathcal{H}$. This inequality is one of the basic analytic tools used in discussing the convergence of sequences of vectors (see the short discussion in Section 2.5.4).

7.3.3 The Generalised Uncertainty Relations

Now we can use the Schwarz inequality to derive an uncertainty relation for any pair of non-commuting operators.

Theorem

If O_1 and O_2 are any two observables then

$$\boxed{\triangle_\psi O_1\,\triangle_\psi O_2 \geq \frac{1}{2}\left|\langle\vec{\psi},[\,\widehat{O}_1,\widehat{O}_2\,]\vec{\psi}\rangle\right|} \tag{7.54}$$

Proof

If $\widehat{A} := \widehat{O}_1 - \langle O_1\rangle_\psi\,\widehat{\mathbb{1}}$ and $\widehat{B} := \widehat{O}_2 - \langle O_2\rangle_\psi\,\widehat{\mathbb{1}}$, then

$$(\triangle_\psi O_1)^2 = \langle(O_1 - \langle O_1\rangle_\psi)^2\rangle_\psi = \langle A^2\rangle_\psi = \langle\vec{\psi},\widehat{A}^{\,2}\vec{\psi}\rangle = \langle\widehat{A}\vec{\psi},\widehat{A}\vec{\psi}\rangle$$

where the last equality follows from the self-adjointness of $\widehat{A}$. Hence

$$\triangle_\psi O_1 = \|\,\widehat{A}\vec{\psi}\,\|$$

and similarly $\Delta_\psi O_2 = \| \widehat{B}\vec{\psi} \|$. Furthermore, $[\widehat{O}_1, \widehat{O}_2] = [\widehat{A}, \widehat{B}]$ (since the unit operator commutes with everything), and so we want to prove that

$$\| \widehat{A}\vec{\psi} \| \, \| \widehat{B}\vec{\psi} \| \geq \frac{1}{2} |\langle \vec{\psi}, [\widehat{A}, \widehat{B}]\vec{\psi}\rangle|.$$

Now, from the Schwarz inequality we have

$$\| \widehat{A}\vec{\psi} \| \, \| \widehat{B}\vec{\psi} \| \geq |\langle \widehat{A}\vec{\psi}, \widehat{B}\vec{\psi}\rangle| = |\langle \vec{\psi}, \widehat{A}\widehat{B}\vec{\psi}\rangle|$$

where the last equality follows because $\widehat{A}$ is self-adjoint. But, as an identity,

$$\widehat{A}\widehat{B} = \frac{1}{2}[\widehat{A}, \widehat{B}]_+ + \frac{1}{2}[\widehat{A}, \widehat{B}]$$

where $[\widehat{A}, \widehat{B}]_+ := \widehat{A}\widehat{B} + \widehat{B}\widehat{A}$ is the *anticommutator* of the two operators. Hence

$$\Delta_\psi O_1 \, \Delta_\psi O_2 \geq \left| \langle \vec{\psi}, \left(\frac{1}{2}[\widehat{A}, \widehat{B}]_+ + \frac{1}{2}[\widehat{A}, \widehat{B}]\right)\vec{\psi}\rangle \right|.$$

However, because $\widehat{A}$ and $\widehat{B}$ are self-adjoint it follows that the number $\langle \vec{\psi}, [\widehat{A}, \widehat{B}]_+\vec{\psi}\rangle$ is real, and $\langle \vec{\psi}, [\widehat{A}, \widehat{B}]\vec{\psi}\rangle$ is imaginary [Exercise]. But, for any real numbers a and b we have $|a + ib|^2 = |a|^2 + |b|^2$, and hence

$$\Delta_\psi O_1 \, \Delta_\psi O_2 \geq \frac{1}{2}\left\{ |\langle \vec{\psi}, [\widehat{A}, \widehat{B}]_+\vec{\psi}\rangle|^2 + |\langle \vec{\psi}, [\widehat{O}_1, \widehat{O}_2]\vec{\psi}\rangle|^2 \right\}^{\frac{1}{2}} \tag{7.55}$$

$$\geq \frac{1}{2}\left| \langle \vec{\psi}, [\widehat{O}_1, \widehat{O}_2]\vec{\psi}\rangle \right|, \tag{7.56}$$

which proves the result. **QED**

Comments

1. The inequality Eq. (7.55) is slightly stronger than the final form Eq. (7.56) although the latter is the most commonly quoted version.

2. There is a natural extension of the generalised uncertainty relations to include the situation in which the state is a density matrix $\widehat{\rho}$.

3. A particular example arises in wave mechanics, where $[\widehat{x}, \widehat{p}] = i\hbar$. In this case Eq. (7.56) gives[5] the familiar uncertainty relations

$$\Delta_\psi x \, \Delta_\psi p \geq \frac{\hbar}{2} \tag{7.57}$$

---

[5]Strictly speaking this is not quite true. Both $\widehat{x}$ and $\widehat{p}$ are unbounded operators and are not defined on the whole Hilbert space $L^2(\mathbb{R})$. The proof of the theorem has to been tightened up in this case.

for all states $\vec{\psi}$.

4. The significance of the state dependence of $\triangle_\psi A$ becomes more apparent if the generalised uncertainty relations are applied to a pair of operators with a non-vanishing commutator that is not just a multiple of the unit operator. A good example is the angular-momentum relation $[\widehat{J}_x, \widehat{J}_y] = i\hbar \widehat{J}_z$, for which Eq. (7.56) gives

$$\triangle_\psi J_x \, \triangle_\psi J_y \geq \frac{\hbar}{2} \langle J_z \rangle_\psi . \tag{7.58}$$

Note that, in states $\vec{\psi}$ for which $\langle J_z \rangle_\psi = 0$, there is no lower bound on the size of the product $\triangle_\psi J_x \, \triangle_\psi J_y$.

7.3.4 A Simple Example

Let us illustrate the generalised uncertainty relations with the aid of a simple model we considered earlier.

Problem

Consider the model dynamical system with Hamiltonian $\widehat{H} = \widehat{\mathbb{1}} + \alpha\sigma_y$ that we discussed in Section 6.4.2.

(i) What is the expected result of S_x in the state $|\psi_t\rangle$ given that $|\psi_0\rangle = \binom{1}{0}$?

(ii) What is the uncertainty $\triangle S_x$ in this state?

(iii) Compute the commutator $[\widehat{S}_x, \widehat{S}_y]$, and hence illustrate the way in which the generalized uncertainty relations are satisfied for the observables S_x and S_y at an arbitrary time t.

Answer

(i) Since $|\psi_t\rangle$ is normalised, the expected result of measuring S_x is $\langle S_x \rangle = \langle \psi_t | \widehat{S}_x | \psi_t \rangle$ where

$$|\psi_t\rangle = e^{-it/\hbar} \begin{pmatrix} \cos \frac{\alpha t}{\hbar} \\ \sin \frac{\alpha t}{\hbar} \end{pmatrix} .$$

The phase factors cancel to give

$$\begin{aligned}
\langle S_x \rangle &= \frac{\hbar}{2}\left\langle \begin{pmatrix} \cos\frac{\alpha t}{\hbar} \\ \sin\frac{\alpha t}{\hbar} \end{pmatrix}, \begin{pmatrix} 0 & 1 \\ 1 & 0 \end{pmatrix} \begin{pmatrix} \cos\frac{\alpha t}{\hbar} \\ \sin\frac{\alpha t}{\hbar} \end{pmatrix} \right\rangle \\
&= \frac{\hbar}{2}\left\langle \begin{pmatrix} \cos\frac{\alpha t}{\hbar} \\ \sin\frac{\alpha t}{\hbar} \end{pmatrix}, \begin{pmatrix} \sin\frac{\alpha t}{\hbar} \\ \cos\frac{\alpha t}{\hbar} \end{pmatrix} \right\rangle \\
&= \frac{\hbar}{2} 2\cos\frac{\alpha t}{\hbar}\sin\frac{\alpha t}{\hbar} = \frac{\hbar}{2}\sin\frac{2\alpha t}{\hbar}. \qquad (7.59)
\end{aligned}$$

(ii) The uncertainty is $(\triangle S_x)^2 = \langle S_x^2 \rangle - \langle S_x \rangle^2$. Now $\widehat{S}_x^2 = \frac{\hbar^2}{4}\left(\begin{smallmatrix} 1 & 0 \\ 0 & 1 \end{smallmatrix}\right)$, and hence $\langle S_x^2 \rangle = \frac{\hbar^2}{4}$. Combining this with Eq. (7.59), we get the result

$$\triangle S_x = \frac{\hbar}{2}\left(1 - \sin^2\frac{2\alpha t}{\hbar}\right)^{\frac{1}{2}} = \frac{\hbar}{2}\left|\cos\frac{2\alpha t}{\hbar}\right|. \qquad (7.60)$$

(iii) The commutator is

$$\begin{aligned}
[\,\widehat{S}_x, \widehat{S}_y\,] &= \frac{\hbar^2}{4}\left[\begin{pmatrix} 0 & 1 \\ 1 & 0 \end{pmatrix}, \begin{pmatrix} 0 & -i \\ i & 0 \end{pmatrix} \right] \\
&= \frac{\hbar^2}{4}\left\{ \begin{pmatrix} 0 & 1 \\ 1 & 0 \end{pmatrix}\begin{pmatrix} 0 & -i \\ i & 0 \end{pmatrix} - \begin{pmatrix} 0 & -i \\ i & 0 \end{pmatrix}\begin{pmatrix} 0 & 1 \\ 1 & 0 \end{pmatrix} \right\} \\
&= \frac{\hbar^2}{4}\left\{ \begin{pmatrix} i & 0 \\ 0 & -i \end{pmatrix} - \begin{pmatrix} -i & 0 \\ 0 & i \end{pmatrix} \right\} = \frac{\hbar^2}{2}\begin{pmatrix} i & 0 \\ 0 & -i \end{pmatrix}
\end{aligned}$$

which shows that $[\,\widehat{S}_x, \widehat{S}_y\,] = i\hbar\widehat{S}_z$.

The generalised uncertainty relation is

$$\triangle S_x \,\triangle S_y \geq \frac{1}{2}|\langle\psi_t|\,[\,\widehat{S}_x, \widehat{S}_y\,]\,|\psi_t\rangle| = \frac{\hbar}{2}|\langle\psi_t|\,\widehat{S}_z\,|\psi_t\rangle| \qquad (7.61)$$

and we must compute both sides. We have already calculated $\triangle S_x$, and therefore the first step is to calculate $\triangle S_y$. Now

$$\begin{aligned}
\langle S_y \rangle &= \frac{\hbar}{2}\left\langle \begin{pmatrix} \cos\frac{\alpha t}{\hbar} \\ \sin\frac{\alpha t}{\hbar} \end{pmatrix}, \begin{pmatrix} 0 & -i \\ i & 0 \end{pmatrix} \begin{pmatrix} \cos\frac{\alpha t}{\hbar} \\ \sin\frac{\alpha t}{\hbar} \end{pmatrix} \right\rangle \\
&= \frac{\hbar}{2}\left\langle \begin{pmatrix} \cos\frac{\alpha t}{\hbar} \\ \sin\frac{\alpha t}{\hbar} \end{pmatrix}, \begin{pmatrix} -i\sin\frac{\alpha t}{\hbar} \\ i\cos\frac{\alpha t}{\hbar} \end{pmatrix} \right\rangle = 0
\end{aligned}$$

and so $(\triangle S_y)^2 = \langle S_y^2 \rangle$. However, $\widehat{S}_y^2 = \frac{\hbar^2}{4}\left(\begin{smallmatrix}1&0\\0&1\end{smallmatrix}\right)$, and hence $\langle S_y^2 \rangle = \frac{\hbar^2}{4}$, which shows that $\triangle S_y = \frac{\hbar}{2}$. Thus, using the result Eq. (7.60), the left hand side of Eq. (7.61) is

$$\triangle S_x \, \triangle S_y = \frac{\hbar^2}{4}\left| \cos \frac{2\alpha t}{\hbar} \right|.$$

On the other hand, the right hand side of Eq. (7.61) is

$$\begin{aligned}
&\frac{\hbar}{2}\left|\left\langle \begin{pmatrix} \cos \frac{\alpha t}{\hbar} \\ \sin \frac{\alpha t}{\hbar} \end{pmatrix}, \frac{\hbar}{2}\begin{pmatrix} 1 & 0 \\ 0 & -1 \end{pmatrix} \begin{pmatrix} \cos \frac{\alpha t}{\hbar} \\ \sin \frac{\alpha t}{\hbar} \end{pmatrix} \right\rangle\right| \\
&= \frac{\hbar^2}{4}\left|\left\langle \begin{pmatrix} \cos \frac{\alpha t}{\hbar} \\ \sin \frac{\alpha t}{\hbar} \end{pmatrix}, \begin{pmatrix} \cos \frac{\alpha t}{\hbar} \\ -\sin \frac{\alpha t}{\hbar} \end{pmatrix} \right\rangle\right| = \frac{\hbar^2}{4}\left|\left(\cos^2 \frac{\alpha t}{\hbar} - \sin^2 \frac{\alpha t}{\hbar}\right)\right| \\
&= \frac{\hbar^2}{4}\left| \cos \frac{2\alpha t}{\hbar} \right|
\end{aligned}$$

so the uncertainty relation is satisfied (just!) for all time.

7.3.5 Some Conceptual Issues

It is important to note that, in a relative-frequency interpretation of quantum theory, an uncertainty relation like Eq. (7.57) has nothing to do with the viability of performing 'simultaneous' measurements! Rather, it refers to the *statistical* spread in the results of making repeated measurements of x and p on large numbers of identically-prepared systems, and this does *not* require that both x and p be measured at the same time. If desired, the inequality Eq. (7.56) can be read as a fundamental limitation on the possibility of *preparing* a quantum state $\vec{\psi}$ to have statistical dispersions that violate the inequality.

Thus, in such approaches to quantum theory, $\triangle_\psi A$ denotes the limiting dispersion of results in a large series of repeated measurements; as such, it is relatively unproblematic. However, in Chapter 5 we referred briefly to realist attempts to assign properties to individual systems and to move away from a relative-frequency interpretation of probability. The meaning ascribed to $\triangle_\psi A$ gives a good insight into the overall structure of an interpretative framework of this type.

The central problem can be viewed from a variety of perspectives, one of which is the meaning that is to be given to a probability statement when applied to a single event. In particular, the 'uncertainty' $\Delta_\psi A$ has to have some such interpretation. Different physicists have responded to this situation in different ways. In his early writings, Heisenberg regarded $\Delta_\psi A$ as a measure of an *epistemological* uncertainty in the *possessed* value of the quantity A. Such a view is still sometimes promulgated today, often accompanied with vague remarks about a measurement of an observable producing an 'uncontrollable disturbance' in its value.

However, the use of such language is very misleading. The central position of standard quantum theory is not that a quantity like A has a value which we happen not to know, but rather that, in a typical quantum state, it is not meaningful to say that A possesses any value at all. And that which does not exist, can not be disturbed; uncontrollably or otherwise. If one still insists on talking about $\Delta_\psi A$ as a measure of a lack of knowledge, it must be appreciated that the word 'knowledge' is being used in a rather odd way, and this can lead to considerable confusion.

Niels Bohr himself was not very keen on Heisenberg's view, and developed the idea that an expression like $\Delta_\psi A$ refers to the extent to which the classical *concepts* employed in describing the observable A are not applicable at the quantum level.

Yet another approach is to adopt the idea mentioned briefly in Chapter 5 in which the probabilistic predictions of quantum theory are deemed to refer to the *latency* (or *potentiality*, or *propensity*) with which the particular values of an observable are possessed by an individual quantum system. A quantity like $\Delta_\psi A$ is then to be viewed as a numerical measure of the extent to which the property A is *not* possessed by the system.

Schemes of this type are very interesting, and suggest strongly that fundamental new concepts are needed to describe the quantum world. However, it is not easy to give a *quantitative* meaning to $\Delta_\psi A$ using categories like 'latency', or 'inapplicability of classical concepts', and in practice one tends to fall back on the relative-frequency interpretation for this purpose. But to proceed any further with this type of analysis requires a detailed look at some of the general conceptual issues in quantum theory: a task to which we now turn.

Chapter 8

CONCEPTUAL ISSUES IN QUANTUM THEORY

8.1 A Quaternity of Problems

The discussion in Section 6.4 makes time development in quantum theory look much the same as that in classical physics: the trajectory of the system in state space is determined uniquely by the dynamical equations once the state at some initial time has been specified. However, in quantum theory another—and quite different—change of state can occur, and that is when measurements are made. This is the so-called *reduction of the state vector*, which is one of the four central problems around which most discussions of the interpretation of quantum theory take place. Within either a pragmatic or a strict instrumentalist approach to quantum theory, one can just about avoid these issues with a clear conscience, but major difficulties arise if one tries to move towards a more realist position, in which states, properties, or both, are posited of individual systems.

The quaternity of fundamental problems is:

1. the meaning of probability;

2. the role of measurement;

3. the reduction of the state vector;

4. quantum entanglement.

The first problem is an old one: what is the meaning of the predictions of a theory that are probabilistic in nature? The second concerns the fundamental role in quantum theory that is apparently played by the concept of 'measurement': is this really necessary, or can we get back to a more realist view in which a measurement merely reveals what was actually the case prior to the measurement? State-vector reduction is a problematic change in the state vector associated with a certain type of measurement: is this, too, a fundamental effect, or can it be viewed as a pragmatic approximation to a Schrödinger-equation evolution for the combined system of quantum object plus measuring device? The final member of the quaternity—quantum entanglement—is perhaps the most challenging of all, with its implication that a quantum whole is not the sum of its parts; or, at least, certainly not in the sense of classical physics.

These four problems are tightly linked together, with each depending strongly on the other three. The explicit manifestation, and general significance, of each varies greatly according to the overall interpretative scheme that is being adopted. Indeed, discussion of these problems can play a major role in illuminating the fundamental differences between the various interpretations of quantum theory, especially in regard to the question of the extent to which a 'realist' view can be taken of properties and states. Each problem has an analogue in the context of classical physics, and this will be discussed briefly in what follows. However, the basic mathematical structure and conceptual framework of classical physics (discussed in Section 4.3) is such that each such analogous problem has a clear resolution, and in a way that is totally consistent with a realist philosophy of physics. As we shall see, the situation in quantum theory is very different.

8.2 The Meaning of Probability

Let us start with the idea of *probability*. Philosophers and probability theorists have argued extensively about the meaning of this concept, even when applied to such mundane acts as tossing a coin, betting on a horse, or (distinctly less mundanely) making a prediction of tomorrow's weather or the spreading of an epidemic. The use of probability in physics is not peculiar to quantum theory; indeed, the ideas of classical statistical

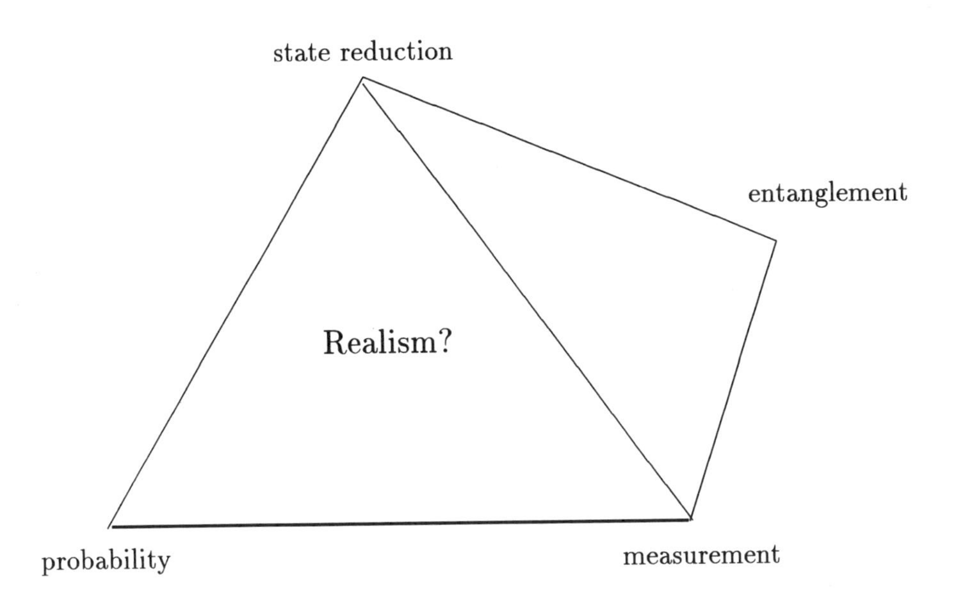

Figure 8.1: A quaternity of problems

physics were well established in the nineteenth century. The most common meaning attached to probability in classical physics is an epistemic one. Thus probabilities measure the extent to which an (idealised)[1] observer *knows* the properties of a system. Our ignorance in this matter arises from the great complexity of the system (for example, a box of gas) but the underlying assumption is that, at least in principle, this uncertainty can be made arbitrarily small with the aid of more precise measurements—or a more detailed theoretical description—of the actual properties of the system.

This position is captured nicely by the mathematical representation of probabilities as measures on a classical state space $\mathcal{S}$, as was discussed briefly in Section 4.3 in the context of Eq. (4.11). In using Eq. (4.11) it is assumed that, at any given time, the physical system *has* a unique state

[1]'Idealised' in the sense that when we talk of 'knowledge' in the context of physics it is usually meant in an intersubjective sense: we are not concerned with the details of individual human psychology!

$s \in \mathcal{S}$ even though we do not know exactly what it is: the number assigned by the probability measure to any subset $W \subset \mathcal{S}$ is then a quantification of the probability that the state is, in fact, in W. A key property of this picture is the existence of probability measures on $\mathcal{S}$ that specify the actual state (and therefore the value of all physical quantities) as accurately as one may wish. In particular, measures exist that specify the state *exactly*. Namely, each $s \in \mathcal{S}$ gives rise to the special 'maximal information' measure μ_s defined by $\mu_s(W) := 1$ if $s \in W$, and $\mu_s(W) := 0$ otherwise.

However, unless hidden variables are posited, the situation in quantum theory is very different. The analogue of a classical probability measure is a density matrix on the Hilbert space $\mathcal{H}$, and the special density matrices $P_{|\psi\rangle} := |\psi\rangle\langle\psi|$ (equivalently, the vectors in $\mathcal{H}$) are the analogues of the maximal-information measures. But now, even if we know that the quantum state is a vector $|\psi\rangle$ (*i.e.,* there is no need to use a non-trivial density matrix), the predictions of the theory are still intrinsically probabilistic in nature. In particular, there are *no* underlying microstates of whose precise values we are ignorant.

If taken seriously, such a view of the probabilistic structure in quantum theory entails a radical departure from the philosophical position of classical physics. The rejection of an epistemic interpretation means that the probabilistic predictions of quantum theory are frequently understood in terms of the quasi-instrumentalist, pragmatic perspective we adopted in the statement of the four Rules. Thus the theory predicts the *relative frequency* of different results if an observable is measured on a large number (ideally, infinite) of identically-prepared systems, but it says nothing about properties of an *individual* system in anything other than this particular probabilistic sense. Indeed, a committed anti-realist may insist that it is not consistent to talk at all of individual entities possessing values for most physical quantities. As John Bell has put it, the *beables* of classical physics are replaced by *observables*, and the verb 'to be' becomes 'to be measurable' (Bell 1987).

If no meaning is attributed to the value of a physical quantity other than as a result of a measurement, then the concept of 'measurement' is bound to play a fundamental role in the formulation of the theory. In the traditional view, dating back to Bohr (for a general discussion, see Scheibe (1973)), the measuring device is located firmly in an external, classical world, thereby opening an irreducible gap between the quantum system and the instrument of observation. This dualism contrasts sharply with the situation in classical physics where all aspects of the physical world are deemed to be describable by the same set of physical laws; in particular, this includes the interaction between an apparatus and the system it is measuring. The issue of dualism cannot be avoided in quantum theory since even the most ardent supporter of the traditional view must agree that real measuring devices are composed of atoms, which certainly require a quantum-theoretical description.

The pragmatic approach to quantum theory is harmless enough when regarded as a fall-back position that does not rule out more adventurous developments in either the technical or the conceptual content of the theory. However, if the pragmatic approach is taken as fundamental, then the implicit philosophical position is strongly instrumentalist in tone. Quantum theory is seen as a set of prescriptions that churns out useful results, but which gives no direct picture of (or assigns any meaning to) the reality that is assumed by most scientists to lie beneath their observations. Many of the conceptual problems are certainly sidestepped by this procedure, but the price paid is an unequivocal anti-realism, and many physicists find it hard to believe that the resulting package is as complete a description of the physical world as its proponents sometimes claim.

Another option is to go along with the realist tendency to ascribe states to individual systems but to forego an epistemic interpretation of probability in favour of a more ontological one. For example, as mentioned earlier, Margenau developed an interpretive scheme for quantum theory in which the classical notion of a system 'possessing' certain attributes is replaced by one in which the properties concerned are *latent* (Margenau 1949, McKnight 1952). Thus a measurement of an observable when the state is not one of its eigenstates, converts a latent value into an 'actual' one. Similar ideas were advocated by Heisenberg (1952) in his later work in which he drew analogies with the old Aristotelian notion of *potentiality*. Popper's concept of *propensity* is another example of this type of idea (Popper 1956). The probability of an event has then to be regarded as some sort of numerical measure of the latency/potentiality/propensity that the particular result will occur.

Note that, in interpretations of this type, the idea of measurement still plays a fundamental role, namely that of inducing a transition from potentiality to actuality. Hence more is needed to get to the situation in which 'measurement' is truly just a name for a particular type of interaction, and one that plays no basic part in the formulation of the theory. In particular, there is still the problem that a real measuring device is built from distinctly quantum-mechanical atoms, and should be capable of being described, therefore, as a *bona fide* quantum system in its own right.

Thus quantum theory is obliged to address two key questions: (i) when is an interaction between two systems to count as a measurement by one of a property of the other?; and (ii) what happens if an attempt is made to restore a degree of unity by describing the measurement process in quantum-theoretical terms, including the internal structure of the measurement device itself? This latter problem is particularly relevant in the context of quantum cosmology, where attempts are made to describe the entire material universe as a single quantum system. As we shall see, there is no universally accepted answer to either of these questions.

8.3 Reduction of the State Vector

8.3.1 Mathematical Aspects

Even on its own, self-limited terms, the pragmatic, relative-frequency approach to quantum theory encounters questions that need to be answered, one of the most important of which is what it means to say, and how it can be ensured, that the individual systems on which the repeated measurements are to be made are all in the 'same'[2] state immediately before the measurement. This crucial problem of *state preparation* is closely related to the idea of a reduction of the state vector.

A measurement is an operation on a system that probes the quantum state immediately *before* the measurement is made, and which yields a definite, and recordable, number concerning a specific observable quantity.

[2]The language is inevitably ambiguous in so far as the pragmatist is uncommitted about whether a 'state' truly refers to a *collection* of copies of a system, or whether it can be consistently thought of as something that is attached to an *individual* system. However, that should not be allowed to obscure the important issue, which is how to prepare quantum systems so that repeated measurements will yield the appropriate results for some specific state vector $\vec{\psi}$.

This is what is implied in the statement of Rule 3. On the other hand, state preparation is an operation whose aim is to force the system (or ensemble[3] of such, if one prefers that view on the referent of a state) to be in some specified state immediately *after* the operation.

It is important to keep separate the ideas of measurement and state preparation. A measurement frequently destroys the system (for example, when a photon activates a Geiger counter, it is absorbed in the process) or else renders it of no further physical interest so far as the measurement in question is concerned (for example, when an electron hits a photographic plate, it is lost among the emulsion's constituent electrons). However, this is of no relevance to the probabilistic predictions of the measurement results. On the other hand, a central feature of state preparation is that, not only is the system not destroyed, but it—or 'ensemble' of such—is left in a particular state. In practice, the preparation of a particular state might involve a special type of measurement, but not all acts of preparation are measurements in the traditional sense.

The notion of state preparation is closely connected with the idea of state reduction. Let A be an observable quantity whose associated self-adjoint operator $\widehat{A}$ has eigenvalues $\{a_1, a_2, \ldots, a_M\}$ that are discrete and non-degenerate. An operation on a collection $\mathcal{E}$ of physical systems is said to be a *state preparation for A* if it leads to a partitioning of $\mathcal{E}$ into subsets $\mathcal{E}_m$, such that for each $m = 1, 2, \ldots, M$, an immediate post-preparation measurement of A can be guaranteed to give the result a_m on each system in the subset $\mathcal{E}_m$. If the operation is also a *bona fide* measurement, then, following Pauli, it is said to be an *ideal measurement* of A.

As a simple example, consider a beam of Silver atoms sent into a Stern–Gerlach device whose magnet is aligned along the z-axis, and suppose this device is followed by a pair of filters, one of which selects the upper beam, and the other of which selects the lower beam. Then any measurements of S_z on the atoms that pass through the upper (respectively lower) filter, will necessarily yield the value $\frac{+\hbar}{2}$ (respectively $\frac{-\hbar}{2}$). Note that this particular state preparation is *not* a measurement; at least, not in the sense that the

[3]Recall my earlier cautionary remark: the use of the word 'ensemble' is *not* meant to imply that physical quantities *have* values that are distributed in an unknown way among the elements in the ensemble. It is more a code word to remind us that, in the pragmatic approach, or full instrumentalist interpretation, the predictions of the theory concern only the spread of the results of repeated measurements.

result is recorded before the subsequent S_z measurement is made.

Now suppose a selection is made by keeping only those systems that lie in a particular subset $\mathcal{E}_m$. Then measurements of A on the systems in $\mathcal{E}_m$ will necessarily yield the result a_m. However, as discussed in section 5.1, the only state that gives a probability-one[4] prediction is an eigenvector of the operator concerned, and—if the eigenvalue is non-degenerate (as we are assuming)—this is unique up to an irrelevant multiplicative factor. Hence we have succeeded in producing a collection of systems (*i.e.*, the set $\mathcal{E}_m$) whose quantum state is guaranteed to be the eigenvector $|a_m\rangle$, *i.e.*, this is the vector that gives the correct probabilistic results of any further measurements made on the systems in $\mathcal{E}_m$. This is the basic way in which quantum states are prepared.

If such a state preparation of A is performed on a collection $\mathcal{E}$ of systems that has previously[5] been prepared to be in a particular state $|\psi\rangle$, then the selection of the sub-collection means that the state $|\psi\rangle$ for $\mathcal{E}$ is replaced by the state $|a_m\rangle$ for $\mathcal{E}_m$. In the approaches to quantum theory that use the relative-frequency interpretation of probability, this transition

$$|\psi\rangle \mapsto |a_m\rangle \tag{8.1}$$

is called a *reduction of the state vector*. However, note that this phrase is often used also within the framework of interpretations in which a state is unequivocally associated with a *single* system. In this case, it is assumed that a single measurement on the system will 'cause' a reduction of the state vector. The existence of such a process is usually presented as a *postulate* (known as von Neumann's 'projection postulate') rather than, as above, something that can be derived from the existing formalism: see below for further discussion.

Returning to the relative-frequency approach, suppose now that the whole collection $\mathcal{E}$ is kept in the sense that the state preparation for A is used to partition $\mathcal{E}$ into subsets $\mathcal{E}_m$, $m = 1, 2, \ldots, M$, but no selection is then made. What state describes the probabilistic distributions of the results of any subsequent observations on the systems in $\mathcal{E}$?

---

[4]I am making the physicist's standard assumption that an event *necessarily* occurring is synonymous with it occurring with probability one.

[5]Throughout this section I am assuming that the time-intervals between sequences of measurements or preparations are sufficiently small that there is no need to consider the internal time evolution associated with the Schrödinger equation.

The answer depends on what type of state preparation was performed. For example, as Wigner (1963) pointed out, in the case of the Stern–Gerlach device it is theoretically feasible to recombine the beams with the aid of a second device aligned along the same axis as the first, in which case it is effectively as if nothing has happened, and we get back the original state $|\psi\rangle$.

However, if the initial preparation involved ideal measurements, then the option arises of 'throwing away' the results, or simply taking no notice of them in the first place. This means we no longer know to which subset $\mathcal{E}_m$ any particular copy of the system belongs: all that can be said is that the probability that any randomly chosen copy belongs to a particular subset $\mathcal{E}_m$ is equal to the probability $|\langle a_m|\psi\rangle|^2$ that applied *before* the ideal measurement of A was made, since this describes correctly the fraction[6] of copies of the system that lie in the subset $\mathcal{E}_m$ of $\mathcal{E}$. We also know that the appropriate state for making further predictions for the subset $\mathcal{E}_m$ is $|a_m\rangle$.

Thus, if A is measured ideally, but *no* selection is made and the results are not kept, the state of the collection $\mathcal{E}$ that correctly predicts the results of any subsequent observations is the *mixed* state $(|a_1\rangle, |a_2\rangle, \ldots, |a_M\rangle; |\langle a_1|\psi\rangle|^2, |\langle a_2|\psi\rangle|^2, \ldots, |\langle a_M|\psi\rangle|^2)$, whose associated density matrix is

$$\widehat{\rho}_{\text{red}} := \sum_{m=1}^{M} |\langle a_m|\psi\rangle|^2 \, \widehat{P}_{|a_m\rangle}. \tag{8.2}$$

Note that, immediately *before* the ideal measurement of A, the numbers $|\langle a_m|\psi\rangle|^2$ represent the quantum probabilities of what will be found. On the other hand, *after* the measurement, they are to be interpreted in a more conventional way as classical probabilities that quantify our ignorance of which vector state the system is in; or, more precisely, of the subset of $\mathcal{E}$ to which a particular copy belongs. The phrase 'reduction of the state vector' is also used to denote this transition

$$|\psi\rangle \mapsto \widehat{\rho}_{\text{red}} = \sum_{m=1}^{M} |\langle a_m|\psi\rangle|^2 \, \widehat{P}_{|a_m\rangle} \tag{8.3}$$

from the vector state $|\psi\rangle$ to the density matrix state $\widehat{\rho}_{\text{red}}$.

---

[6]Strictly speaking, for any finite set $\mathcal{E}_m$, this fraction only approximates the quantum probability.

Comments

1. If the spectrum of $\widehat{A}$ is degenerate the situation is more complicated since, from the knowledge that a measurement of A will necessarily yield a particular eigenvalue a_m, it is possible to deduce only that the new state is some linear combination of the eigenvectors associated with that eigenvalue.

Lüders (1951) extended von Neumann's projection postulate to such cases with the claim that, if an ideal measurement of A gives the result a_m, then the state appropriate for the subset of systems $\mathcal{E}_m$ after the measurement is $\widehat{P}_m \left|\psi\right\rangle$, where $\widehat{P}_m := \sum_{j=1}^{d(m)} |a_m, j\rangle\langle a_m, j|$ is the operator that projects onto the subspace of eigenvectors of $\widehat{A}$ with eigenvalue a_m. The normalised form of this vector is $\widehat{P}_m \left|\psi\right\rangle / \parallel \widehat{P}_m \left|\psi\right\rangle \parallel = \widehat{P}_m \left|\psi\right\rangle / \langle\psi| \widehat{P}_m \left|\psi\right\rangle^{\frac{1}{2}}$, and hence the reduction is

$$\left|\psi\right\rangle \mapsto \widehat{P}_m \left|\psi\right\rangle / \langle\psi| \widehat{P}_m \left|\psi\right\rangle^{\frac{1}{2}}. \tag{8.4}$$

This is illustrated in Figure 8.2 for an example in which the eigenvalue a_1 is two-fold degenerate, and the eigenvalue a_2 is non-degenerate. The projection is onto the two-dimensional eigenspace associated with the eigenvalue a_1.

Note that, if the eigenvalues of $\widehat{A}$ are nondegenerate, we have $\widehat{P}_m = |a_m\rangle\langle a_m|$, and hence the right hand side of Eq. (8.4) becomes $\frac{\langle a_m|\psi\rangle}{|\langle\psi|a_m\rangle|} |a_m\rangle$ which, since $\frac{\langle a_m|\psi\rangle}{|\langle\psi|a_m\rangle|}$ is a complex number of modulus 1, essentially reproduces Eq. (8.1), as it should.

2. The status of Lüders' projection postulate has been much debated in the past. It is perhaps best regarded as a definition of what is meant by an 'ideal measurement' in the case of a degenerate eigenvalue. Clearly it bears on the question about what would happen if simultaneous ideal measurements were to be made of a complete set of commuting observables that includes A.

3. To find the result that corresponds to Eq. (8.4) when no selection is made, note that, since the probability of getting the result a_m is $\text{Prob}(A = a_m; \left|\psi\right\rangle) = \langle\psi| \widehat{P}_m \left|\psi\right\rangle = \text{tr}\,(\widehat{P}_{|\psi\rangle}\widehat{P}_m)$, the analogue of Eq. (8.3) is

$$\left|\psi\right\rangle \mapsto \widehat{\rho}_{\text{red}} = \sum_{m=1}^{M} \langle\psi| \widehat{P}_m \left|\psi\right\rangle \frac{\widehat{P}_m \left|\psi\right\rangle\langle\psi| \widehat{P}_m}{\langle\psi| \widehat{P}_m \left|\psi\right\rangle} = \sum_{m=1}^{M} \widehat{P}_m \widehat{P}_{|\psi\rangle} \widehat{P}_m \tag{8.5}$$

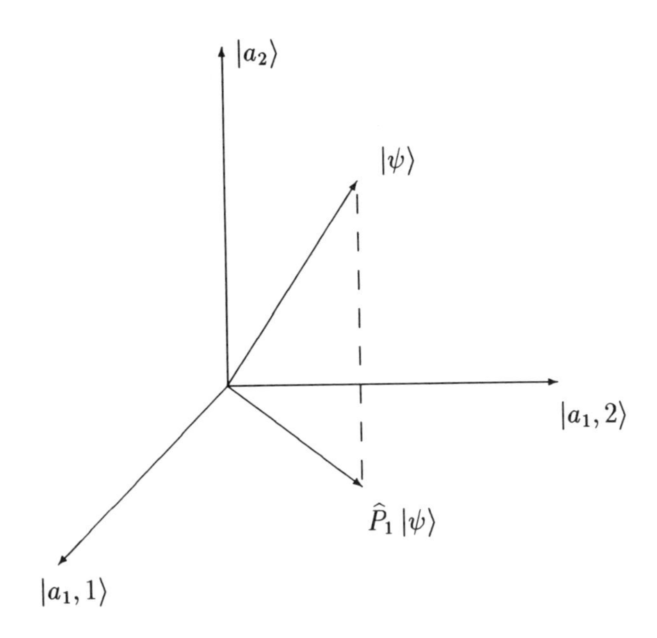

Figure 8.2: The Lüders Projection

where $\widehat{P}_{|\psi\rangle} := |\psi\rangle\langle\psi|$.

4. A more complicated example of reduction is when the initial state is a *mixed* state, represented by a density matrix $\widehat{\rho}$. To gain some insight into what happens here, consider first the special case when $\widehat{\rho}$ is just the density matrix form of a vector state $|\psi\rangle$, *i.e.*, $\widehat{\rho} = \widehat{P}_{|\psi\rangle}$. The reduction Eq. (8.4) can then be rewritten as the transformation of density matrices

$$\widehat{P}_{|\psi\rangle} \mapsto \frac{\widehat{P}_m \,|\psi\rangle\langle\psi|\, \widehat{P}_m}{\langle\psi|\, \widehat{P}_m \,|\psi\rangle} = \frac{\widehat{P}_m \widehat{P}_{|\psi\rangle} \widehat{P}_m}{\mathrm{tr}\,(\widehat{P}_m \widehat{P}_{|\psi\rangle})}. \tag{8.6}$$

This expression suggests that if the initial state is *any* mixed state $\widehat{\rho}$, and if a selection is made of those systems for which the result a_m was obtained, then the appropriate state for predicting the results of future measurements on this selected subset of systems is the right hand side of the reduction

$$\widehat{\rho} \mapsto \widehat{\rho}_{\mathrm{red}} := \frac{\widehat{P}_m \widehat{\rho} \widehat{P}_m}{\operatorname{tr}(\widehat{P}_m \widehat{\rho})}. \tag{8.7}$$

5. Finally, if the initial state is a density matrix $\widehat{\rho}$, and if no selection is made, the appropriate reduction is

$$\widehat{\rho} \mapsto \widehat{\rho}_{\mathrm{red}} = \sum_{m=1}^{M} \operatorname{tr}(\widehat{P}_m \widehat{\rho}) \frac{\widehat{P}_m \widehat{\rho} \widehat{P}_m}{\operatorname{tr}(\widehat{P}_m \widehat{\rho})} = \sum_{m=1}^{M} \widehat{P}_m \widehat{\rho} \widehat{P}_m \tag{8.8}$$

where the original quantum probabilities $\operatorname{tr}(\widehat{P}_m \widehat{\rho})$ that the result $A = a_m$ will be obtained now serve as the classical uncertainties of which result was, in fact, obtained. Note that the density-matrix results Eq. (8.7–8.8) for the case when a_m is *non*-degenerate are simply given by writing $\widehat{P}_m$ as $|a_m\rangle\langle a_m|$.

8.3.2 The Role of Conditional Probability

The 'jump' movements associated with the various types of state reduction have been the subject of much debate over the years. They are not particularly problematic within the pragmatic, relative-frequency approach to quantum theory since, as was explained above, they correspond merely to the observer deciding to restrict his or her attention by selecting those systems on which further measurements are to be made.

Thus, for example, in the reduction Eq. (8.1) it can be argued that the state $|\psi\rangle$ concerns predictions of the results of measurements to be made on the original collection $\mathcal{E}$ of copies of the system, whereas the reduced state $|a_m\rangle$ refers to measurements to be made on the subset $\mathcal{E}_m$ of systems obtained by an appropriate state preparation for A. Indeed, as has been remarked several times before, a keen anti-realist may insist that an individual system does not have a state at all: a 'state' has a meaning only in relation to a (possibly idealised) collection of copies of a system on which repeated measurements are to be made. However, the situation is very different for a realist who aspires to associate states with individual systems. He or she may wish to add the projection postulate as one of the basic Rules of quantum theory, and then to view state-reduction as a reflection of an objective change in the system itself. We shall return briefly to this issue at the end of this section.

The 'system-selection' way of looking at state reduction is quite attractive, not least because it appears to tie in nicely with the well-known procedure in classical probability theory whereby the probabilistic predictions of the values of some quantity B change if information is acquired about the actual value of some other quantity A. In other words, the absolute probability $\mathrm{Prob}(B = b_n)$ of getting a specific value b_n for B is replaced with the *conditional* probability $\mathrm{Prob}(B = b_n \mid A = a_m)$ of finding b_n *given* that the value of A is a_m.

To see how this works in classical physics, let A and B be a pair of physical quantities with (for simplicity) discrete[7] sets of possible values $a_1, a_2, \ldots, a_{M_A}$ and $b_1, b_2, \ldots, b_{M_B}$ respectively. The state space $\mathcal{S}$ can be partitioned into disjoint subsets $\mathcal{S}_{A=a_m}$, where a state s lies in $\mathcal{S}_{A=a_m}$ if and only if A has the value a_m in s, *i.e.*, $f_A(s) = a_m$; and similarly there is another partition of $\mathcal{S}$ into subsets $\mathcal{S}_{B=b_n}$ according to the values of B. If the values of physical quantities are distributed probabilistically according to some probability measure μ on $\mathcal{S}$, then the probabilities of A and B having the values a_m and b_n are simply

$$\mathrm{Prob}(A = a_m; \mu) = \mu(\mathcal{S}_{A=a_m}),\ m = 1, 2, \ldots, M_A \tag{8.9}$$

and

$$\mathrm{Prob}(B = b_n; \mu) = \mu(\mathcal{S}_{B=b_n})\ \ n = 1, 2, \ldots, M_B \tag{8.10}$$

respectively.

Now suppose we discover that, in fact, A has a specific value, a_m say. This increases our knowledge since we now know that the state of the system definitely lies in the subset $\mathcal{S}_{A=a_m}$ of $\mathcal{S}$. In effect, this becomes a new state space for use in predicting the values of B and, not surprisingly, the probability measure changes. The new measure $\mu_{A=a_m}$ is defined by specifying its value on any subset W of $\mathcal{S}$ as

$$\mu_{A=a_m}(W) := \frac{\mu(W \bigcap \mathcal{S}_{A=a_m})}{\mu(\mathcal{S}_{A=a_m})} \tag{8.11}$$

[7]There is no real loss of generality in doing this. If a physical quantity takes on a continuous range of possible values one can always divide this range into disjoint 'bins'.

which, it will be noted, correctly satisfies the normalisation condition $\mu_{A=a_m}(\mathcal{S}_{A=a_m}) = 1$. In particular, the conditional probability $\text{Prob}(B = b_n \mid A = a_m; \mu)$ of B having the value b_n given that the value of A is a_m is

$$\text{Prob}(B = b_n \mid A = a_m; \mu) := \mu_{A=a_m}(\mathcal{S}_{B=b_n}) = \frac{\mu(\mathcal{S}_{B=b_n} \cap \mathcal{S}_{A=a_m})}{\mu(\mathcal{S}_{A=a_m})}. \quad (8.12)$$

Note that this is consistent with the intuitive idea that the joint probability of finding $B = b_n$ *and* $A = a_m$ (which is equal to $\mu(\mathcal{S}_{B=b_n} \cap \mathcal{S}_{A=a_m})$) should be equal to the conditional probability of finding $B = b_n$ given that $A = a_m$, multiplied by the absolute probability of finding $A = a_m$.

Note that the transformation of probability measures

$$\mu \mapsto \mu_{A=a_m} \quad (8.13)$$

reflects our discovery that A has a specific value. As such, it can be regarded as a classical analogue of the reductions Eq. (8.1), Eq. (8.4) or Eq. (8.7), *i.e.*, those quantum reductions in which a selection is made using the outcomes of the first measurement.

An important observation about classical conditional probabilities is that the probabilistic predictions of B do *not* change if we add only the information that A has *some* value, but do not specify which. Indeed, this is implicit in what is meant by saying that every physical quantity always 'has' a value; one might reasonably require this to be a consistency condition in any interpretation of a physical theory that is to be labelled 'realist' in relation to properties. The following theorem shows how this expectation is satisfied in classical statistical physics (and, indeed, in all standard probability theory).

Theorem

The conditional probability $\text{Prob}(B = b_n \mid A; \mu)$ that $B = b_n$, given only that A has *some* value, is equal to the absolute probability $\text{Prob}(B = b_n; \mu)$.

Proof

We note first that this conditional probability $\text{Prob}(B = b_n \mid A; \mu)$ is equal to the sum over m of the absolute probability that $A = a_m$ times the conditional probability that $B = b_n$ given that $A = a_m$:

$$\begin{aligned}\operatorname{Prob}(B = b_n \mid A; \mu) &= \sum_{m=1}^{M_A} \operatorname{Prob}(A = a_m; \mu) \operatorname{Prob}(B = b_n \mid A = a_m; \mu) \\ &= \sum_{m=1}^{M_A} \mu(\mathcal{S}_{A=a_m}) \frac{\mu(\mathcal{S}_{B=b_n} \bigcap \mathcal{S}_{A=a_m})}{\mu(\mathcal{S}_{A=a_m})} \\ &= \sum_{m=1}^{M_A} \mu(\mathcal{S}_{B=b_n} \bigcap \mathcal{S}_{A=a_m}). \end{aligned} \tag{8.14}$$

However, for each fixed n, the subsets $\mathcal{S}_{B=b_n} \bigcap \mathcal{S}_{A=a_m}$, $m = 1, 2, \ldots, M_A$, are disjoint; and hence, from the basic additivity property of a probability measure, it follows that, for each $n = 1, 2, \ldots, M_B$,

$$\begin{aligned}\sum_{m=1}^{M_A} \mu\Big(\mathcal{S}_{B=b_n} \bigcap \mathcal{S}_{A=a_m}\Big) &= \mu\Big(\bigcup_{m=1}^{M_A} \big(\mathcal{S}_{B=b_n} \bigcap \mathcal{S}_{A=a_m}\big)\Big) \\ &= \mu\Big(\mathcal{S}_{B=b_n} \bigcap \Big(\bigcup_{m=1}^{M_A} \mathcal{S}_{A=a_m}\Big)\Big) \end{aligned} \tag{8.15}$$

where in the second equality we have used the distributivity law $(A \cap B) \cup (A \cap C) = A \cap (B \cup C)$ for subsets of a set. However, $\bigcup_{m=1}^{M_A} \mathcal{S}_{A=a_m} = \mathcal{S}$, and $\mathcal{S}_{B=b_n} \cap \mathcal{S} = \mathcal{S}_{B=b_n}$. Thus Eq. (8.14) implies that, for each n,

$$\operatorname{Prob}(B = b_n \mid A; \mu) = \mu(\mathcal{S}_{B=b_n}) = \operatorname{Prob}(B = b_n; \mu), \tag{8.16}$$

as required. **QED**

Now consider the analogous calculation in quantum theory. According to the reduction Eq. (8.1), the appropriate quantum state to use if an ideal measurement of A gives the value a_m, is $|a_m\rangle$; and hence the probability of a subsequent measurement of B giving the result b_n is

$$\operatorname{Prob}(B = b_n \mid A = a_m) = |\langle b_n | a_m \rangle|^2 \tag{8.17}$$

which, it should be noted, is independent of $|\psi\rangle$.

This expression looks as if it should be viewed as the quantum analogue of the classical conditional probability Eq. (8.12). However, there is a profound difference between them. To see this, we note that if the initial quantum state is $|\psi\rangle$, the absolute probability for getting the result $A =$

a_m is $\text{Prob}(A = a_m;\ |\psi\rangle) = |\langle a_m|\psi\rangle|^2$, and hence the probability $\text{Prob}(B = b_n \mid A;\ |\psi\rangle)$ that we shall find the particular result $B = b_n$ given only that there *was* a previous ideal measurement of A is

$$\begin{aligned}
\text{Prob}(B = b_n \mid A;\ |\psi\rangle) &= \sum_{m=1}^{M_A} \text{Prob}(B = b_n \mid A = a_m)\text{Prob}(A = a_m;\ |\psi\rangle) \\
&= \sum_{m=1}^{M_A} |\langle b_n|a_m\rangle|^2\ |\langle a_m|\psi\rangle|^2 \\
&= \sum_{m=1}^{M_A} \langle\psi|a_m\rangle\langle a_m|b_n\rangle\langle b_n|a_m\rangle\langle a_m|\psi\rangle. \qquad (8.18)
\end{aligned}$$

Equivalently, we can take the reduced density matrix in Eq. (8.3), which is appropriate for the situation in which no account is taken of the result of the ideal measurement of A, and argue directly that

$$\text{Prob}(B = b_n \mid A;\ |\psi\rangle) = \text{tr}\,(\widehat{\rho}_{red}\widehat{P}_{|b_n\rangle}) = \sum_{m=1}^{M_A} |\langle a_m|\psi\rangle|^2 \text{tr}\,(\widehat{P}_{|a_m\rangle}\widehat{P}_{|b_n\rangle}), \qquad (8.19)$$

which gives the same result.

On the other hand, if B is measured, the absolute probability of getting the result b_n is

$$\text{Prob}(B = b_n;\ |\psi\rangle) = |\langle b_n|\psi\rangle|^2 \equiv \sum_{m_1=1}^{M_A}\sum_{m_2=1}^{M_A} \langle\psi|a_{m_1}\rangle\langle a_{m_1}|b_n\rangle\langle b_n|a_{m_2}\rangle\langle a_{m_2}|\psi\rangle, \qquad (8.20)$$

which is *not* the same as Eq. (8.18).

Thus quantum-mechanical conditional probabilities behave in a very different way from classical ones. In particular, the classical probabilistic distribution of the values of B does *not* depend on the fact that A *has* some value, but the quantum-mechanical predictions of results of measuring B *do* depend on the fact that an ideal *measurement* of A was performed immediately beforehand.

One might argue that this apparently counter-intuitive behaviour of the quantum conditional probability is, in fact, just what should have been expected. As we are always being told, a quantum measurement, even an 'ideal' one, causes an "uncontrollable disturbance of the system", and this is precisely what the first, ideal measurement of A has done. It is not surprising therefore that subsequent measurements of B produce a different probabilistic spread from what would have been obtained had A not been measured. However, the use of such language is misleading in any interpretation of quantum theory that does not give an objective status to properties of an individual system: if there are no possessed properties, how can a measurement disturb them, uncontrollably or otherwise? The most that might be said is that a measurement produces an uncontrollable disturbance in the *potentiality* for different results to be obtained in a later measurement, but this is arguably a rather different matter.

8.3.3 The Problem for a Realist

The discussion above has been within the framework of the pragmatic approach to quantum theory, and applies as it stands to a full-blown instrumentalist interpretation. However, the idea of a reduction of the state vector is often invoked in more realist approaches in which the state vector is deemed to refer to a single system. The reduction is then assumed to occur after a *single* (ideal) measurement, and has nothing to do with system selection in a series of repeated measurements. From this perspective, the overall time development of a state of a single system consists of sharp jumps produced by the act of measurement, separated by periods of deterministic evolution governed by the Schrödinger equation of Rule 4.

The major problem is to understand the origin of these sudden changes in the state. In particular, can they be obtained from the existing quantum formalism, or does the reduction of the state vector have to be added to the general rules of quantum theory as a fundamental postulate? This problem is particularly acute in any approach to quantum theory that aspires to demote 'measurement' from playing a fundamental part in the formulation of the theory. In this case, there is a strong motivation to try to derive the state reduction from the existing formalism; albeit, perhaps, only as an empirically useful approximation to the actual development of the state in time.

The nature of the problem depends in part on the perceived referent of the state. If the state is held to quantify our *knowledge* of the system, then the reduction process is arguably analogous to the conditioning procedure in classical probability in which the addition of extra information about what is actually the case changes our state of knowledge. On the other hand, if the state vector is held to refer to the system itself, then the idea of reduction is frequently tied to the 'uncontrollable disturbance' thesis. This raises the obvious question of the possibility of understanding the nature of this effect in direct physical terms. In particular, what type of interaction serves as an 'ideal measurement'?

One approach to this problem is to ask again about the significance of the fact that actual measuring devices are made of quantum atoms. Is it possible to understand a state reduction as the outcome of some dynamical evolution in which object and apparatus are both regarded as quantum-mechanical systems? Indeed, even within the minimal, pragmatic approach to quantum theory there is good reason for asking what *type* of interaction between two systems is to be regarded as a *bona fide* measurement of one by the other. The concept of measurement plays a fundamental role in the formulation of quantum theory, and therefore deserves to be understood further.

8.4 Quantum Entanglement

8.4.1 The Tensor Product of Two Hilbert Spaces

Any attempt to describe the act of measurement as a *bona fide* physical process quickly encounters the phenomenon of quantum entanglement: the fourth in our list of basic problems, and one that lies at the heart of some of the most peculiar features of quantum theory.

Quantum entanglement stems from the existence of vector states of a composite quantum system (or suitable 'ensemble' of such systems) for which it is *not* possible to assign vector states to the constituent subsystems. Therefore, the first task is to study the state space of a composite system of S_1 plus S_2 given that the state spaces of the subsystems S_1 and S_2 are Hilbert spaces $\mathcal{H}_1$ and $\mathcal{H}_2$ respectively. The analogous situation in classical physics is very clear: the (micro) states of S_1 plus S_2 consist of all pairs (s_1, s_2) where s_1 and s_2 are states of S_1 and S_2 respectively. Thus the state space of S_1 plus S_2 is just the Cartesian product $\mathcal{S}_1 \times \mathcal{S}_2$; in particular, every microstate of a composite system is associated with a unique microstate for each of its constituent subsystems.

Let us start by considering the example of a quantum system consisting of a spin-half particle and a spin-one particle. If we concentrate on the spin features alone (*i.e.*, ignore the degrees of freedom associated with the spatial properties of the particles), the state spaces of the subsystems are $\mathbb{C}^2$ and $\mathbb{C}^3$ respectively, with typical basis sets being the eigenstates $\{|+\frac{1}{2}\rangle, |-\frac{1}{2}\rangle\}$ and $\{|+1\rangle, |0\rangle, |-1\rangle\}$ of the z-components of the appropriate spin operators. A natural physical assumption is that the composite system will include states in which each particle takes on one of its allowed S_z values. There are clearly six such states, which we shall denote as

$$|+\tfrac{1}{2},+1\rangle,\ |+\tfrac{1}{2},0\rangle,\ |+\tfrac{1}{2},-1\rangle,\ |-\tfrac{1}{2},+1\rangle,\ |-\tfrac{1}{2},0\rangle,\ |-\tfrac{1}{2},-1\rangle \quad (8.21)$$

so that, for example, $|+\frac{1}{2},+1\rangle$ is a state in which the first particle has an S_z of $\frac{1}{2}\hbar$, and the second particle has an S_z of $1\hbar$. More precisely, we assume that the z-components of spin for the spin-half and the spin-one particle can be represented on the Hilbert space of the composite system by operators $\widehat{S}_z^{(1)}$ and $\widehat{S}_z^{(2)}$ respectively. Furthermore, these two variables are expected to be simultaneously[8] measurable, so that

$$[\,\widehat{S}_z^{(1)}, \widehat{S}_z^{(2)}\,] = 0. \quad (8.22)$$

Then the states in Eq. (8.21) denote the simultaneous eigenvectors of these operators. Thus, for example, $\widehat{S}_z^{(1)}\,|+\frac{1}{2},+1\rangle = \frac{1}{2}\hbar\,|+\frac{1}{2},+1\rangle$ and $\widehat{S}_z^{(2)}\,|+\frac{1}{2},+1\rangle = \hbar\,|+\frac{1}{2},+1\rangle$.

Since we are dealing with quantum theory, we expect linear combinations of the above states to be allowed (barring some unforeseen superselection rule). Indeed, it seems natural to suppose that the most general state can be written in the form

$$\begin{aligned}|\psi\rangle = \psi_{+\frac{1}{2},+1}\,|+\tfrac{1}{2},+1\rangle + \psi_{+\frac{1}{2},0}\,|+\tfrac{1}{2},0\rangle + \psi_{+\frac{1}{2},-1}\,|+\tfrac{1}{2},-1\rangle +\\ \psi_{-\frac{1}{2},+1}\,|-\tfrac{1}{2},+1\rangle + \psi_{-\frac{1}{2},0}\,|-\tfrac{1}{2},0\rangle + \psi_{-\frac{1}{2},-1}\,|-\tfrac{1}{2},-1\rangle \quad (8.23)\end{aligned}$$

so that the vectors in Eq. (8.21) form a basis set for the Hilbert space of the composite system. Since there are six such vectors, this space is isomorphic to $\mathbb{C}^6$.

[8]But recall our earlier discussion in Section 6.3 of the idea of two variables being 'simultaneously measurable'.

For a general composite system, the relevant mathematical operation takes vectors $\vec{\psi}_1$ and $\vec{\psi}_2$ in Hilbert spaces $\mathcal{H}_1$ and $\mathcal{H}_2$ respectively, and turns them into a vector $\vec{\psi}_1 \otimes \vec{\psi}_2$ in a new Hilbert space $\mathcal{H}_1 \otimes \mathcal{H}_2$ known as the *tensor product* of $\mathcal{H}_1$ and $\mathcal{H}_2$. The quantum state space of the composite system is then the tensor product of the quantum state spaces of its constituent subsystems. Thus, in the example above, the vector $|+\frac{1}{2}, +1\rangle$ denotes the tensor product $|+\frac{1}{2}\rangle \otimes |+1\rangle$; in particular, we expect the tensor product $\mathbb{C}^2 \otimes \mathbb{C}^3$ to be isomorphic to $\mathbb{C}^6$.

The full definition of the tensor product operation is quite complex but, for our purposes, the following remarks will suffice:

(a) The tensor product operation is linear in each 'slot' in the sense that, for all $\alpha, \beta \in \mathbb{C}$, $\vec{\psi}, \vec{\phi} \in \mathcal{H}_1$ and $\chi \in \mathcal{H}_2$,

$$(\alpha\vec{\psi} + \beta\vec{\phi}) \otimes \vec{\chi} = (\alpha\vec{\psi}) \otimes \vec{\chi} + (\beta\vec{\phi}) \otimes \vec{\chi}, \tag{8.24}$$

and, for all $\alpha, \beta \in \mathbb{C}$, $\vec{\psi} \in \mathcal{H}_1$ and $\vec{\phi}, \vec{\chi} \in \mathcal{H}_2$,

$$\vec{\psi} \otimes (\alpha\vec{\phi} + \beta\vec{\chi}) = \vec{\psi} \otimes (\alpha\vec{\phi}) + \vec{\psi} \otimes (\beta\vec{\chi}). \tag{8.25}$$

Furthermore, multiplication by a complex number carries across the tensor product in the sense that

$$\alpha(\vec{\psi} \otimes \vec{\phi}) = (\alpha\vec{\psi}) \otimes \vec{\phi} = \vec{\psi} \otimes (\alpha\vec{\phi}) \tag{8.26}$$

for all $\alpha \in \mathbb{C}$, $\vec{\psi} \in \mathcal{H}_1$ and $\vec{\phi} \in \mathcal{H}_2$.

(b) There exist vectors in $\mathcal{H}_1 \otimes \mathcal{H}_2$ that *cannot* be written as a single product $\vec{\psi} \otimes \vec{\phi}$ for any $\vec{\psi} \in \mathcal{H}_1$ or $\vec{\phi} \in \mathcal{H}_2$. However, every vector in $\mathcal{H}_1 \otimes \mathcal{H}_2$ can be expressed as a *sum* of such product vectors.

In particular, if $\{\vec{e}_1, \vec{e}_2, \ldots, \vec{e}_{N_1}\}$ and $\{\vec{f}_1, \vec{f}_2, \ldots, \vec{f}_{N_2}\}$ are sets of basis vectors for $\mathcal{H}_1$ and $\mathcal{H}_2$ respectively, then a basis for $\mathcal{H}_1 \otimes \mathcal{H}_2$ is the set of vectors $\vec{e}_i \otimes \vec{f}_j$, $i = 1, 2, \ldots, N_1$, $j = 1, 2, \ldots, N_2$. Thus the most general vector $\vec{\psi} \in \mathcal{H}_1 \otimes \mathcal{H}_2$ can be written as

$$\vec{\psi} = \sum_{i=1}^{N_1} \sum_{j=1}^{N_2} \psi_{ij}\, \vec{e}_i \otimes \vec{f}_j. \tag{8.27}$$

In particular, this shows that the dimension of the Hilbert space $\mathcal{H}_1 \otimes \mathcal{H}_2$ is the product of the dimensions of $\mathcal{H}_1$ and $\mathcal{H}_2$ (for this purpose, $\infty \times \infty =$

∞). Thus, as anticipated, the example considered above corresponds to the isomorphism of $\mathbb{C}^2 \otimes \mathbb{C}^3$ with $\mathbb{C}^6$.

(c) The scalar product is defined on product vectors by

$$\langle \vec{\psi}_1 \otimes \vec{\psi}_2, \vec{\phi}_1 \otimes \vec{\phi}_2 \rangle := \langle \vec{\psi}_1, \vec{\phi}_1 \rangle_{\mathcal{H}_1} \langle \vec{\psi}_2, \vec{\phi}_2 \rangle_{\mathcal{H}_2} \tag{8.28}$$

where the scalar products on the right hand side are computed in the indicated Hilbert spaces. This expression is extended to sums of vectors by defining

$$\begin{aligned} &\langle \vec{\psi}_1 \otimes \vec{\psi}_2, (\alpha\vec{\phi}_1 \otimes \vec{\phi}_2 + \beta\vec{\phi}_3 \otimes \vec{\phi}_4) \rangle := \\ &\quad \alpha\langle \vec{\psi}_1, \vec{\phi}_1 \rangle_{\mathcal{H}_1} \langle \vec{\psi}_2, \vec{\phi}_2 \rangle_{\mathcal{H}_2} + \beta\langle \vec{\psi}_1, \vec{\phi}_3 \rangle_{\mathcal{H}_1} \langle \vec{\psi}_2, \vec{\phi}_4 \rangle_{\mathcal{H}_2} \end{aligned} \tag{8.29}$$

and similarly for linear sums in the left hand side of the scalar product.

(d) Tensor products of operators can also be defined. Specifically, if $\widehat{A}_1$ and $\widehat{A}_2$ are operators on $\mathcal{H}_1$ and $\mathcal{H}_2$ respectively, the tensor product $\widehat{A}_1 \otimes \widehat{A}_2$ is defined first on product vectors by

$$(\widehat{A}_1 \otimes \widehat{A}_2)\vec{\psi}_1 \otimes \vec{\psi}_2 := (\widehat{A}_1\vec{\psi}_1) \otimes (\widehat{A}_2\vec{\psi}_2) \tag{8.30}$$

and then extended to sums of product vectors in the obvious linear way. In particular, using the expansion in Eq. (8.27), we have

$$(\widehat{A}_1 \otimes \widehat{A}_2)\vec{\psi} := \sum_{i=1}^{N_1}\sum_{j=1}^{N_2} \psi_{ij}(\widehat{A}_1\vec{e}_i) \otimes (\widehat{A}_2\vec{f}_j). \tag{8.31}$$

Analogously to the case of vectors, there are operators on $\mathcal{H}_1 \otimes \mathcal{H}_2$ that cannot be written in the form $\widehat{A}_1 \otimes \widehat{A}_2$. However, all operators can be expressed as a sum of such product operators.

Comments

1. In Dirac notation, the tensor product of two states $|\psi\rangle$ and $|\phi\rangle$ is written as $|\psi\rangle\,|\phi\rangle$. Thus, in the simple example above, $|+\frac{1}{2}, +1\rangle$ denotes the product state $|+\frac{1}{2}\rangle\,|+1\rangle$.

2. An observable A_1 for system 1 can be represented in the composite system by the operator $\widehat{A}_1 \otimes \widehat{\mathbb{1}}$. Similarly, an observable A_2 for system 2 is represented by $\widehat{\mathbb{1}} \otimes \widehat{A}_2$. Thus, in the example above, $\widehat{S}_z^{(1)}$ is really $\widehat{S}_{z1} \otimes \widehat{\mathbb{1}}$,

and $\widehat{S}_z^{(2)}$ is $\widehat{\mathbb{1}} \otimes \widehat{S}_{z2}$, where $\widehat{S}_{z1}$ and $\widehat{S}_{z2}$ denote the spin-z operators on $\mathbb{C}^2$ and $\mathbb{C}^3$ respectively.

Note that, even if $\widehat{A}_1$ is free of any degeneracy on $\mathcal{H}_1$, the operator $\widehat{A}_1 \otimes \widehat{\mathbb{1}}$ that acts on $\mathcal{H}_1 \otimes \mathcal{H}_2$ is highly degenerate since $(\widehat{A}_1 \otimes \widehat{\mathbb{1}}) \, |a\rangle \, |\phi\rangle = a \, |a\rangle \, |\phi\rangle$ for any vector $|\phi\rangle \in \mathcal{H}_2$.

3. If A_1, A_2 are a pair of observables for systems 1 and 2 respectively, then

$$[\widehat{A}_1 \otimes \widehat{\mathbb{1}}, \widehat{\mathbb{1}} \otimes \widehat{A}_2] = 0 \tag{8.32}$$

which implies that, thought of as referring to the composite system, A_1 and A_2 are trivially compatible (*i.e.*, they are simultaneously measurable in an uncontentious sense). A simple example of this idea is provided by Eq. (8.22).

Furthermore, if $|\psi\rangle$ is a product state $|\psi_1\rangle \, |\psi_2\rangle$ then, assuming for simplicity that $\widehat{A}_1$ and $\widehat{A}_2$ are both non-degenerate, it is easy to show that the probability distributions for these observables viewed as referring to the composite system reproduce those for the constituent systems. More precisely, as compatible observables, $\widehat{A}_1 \otimes \widehat{\mathbb{1}}$ and $\widehat{\mathbb{1}} \otimes \widehat{A}_2$ possess a complete set of simultaneous eigenvectors, which are clearly the states $|a_1\rangle \, |a_2\rangle$ where a_1 and a_2 range over the eigenvalues of $\widehat{A}_1$ and $\widehat{A}_2$ respectively. Thus in a product state we have (*cf.* Eq. (6.32))

$$\begin{aligned} \mathrm{Prob}(A_1 = a_1, A_2 = a_2; \, |\psi_1\rangle \, |\psi_2\rangle) &= \left|(\langle a_1| \, \langle a_2| \,)(\, |\psi_1\rangle \, |\psi_2\rangle)\right|^2 \\ &= |\langle a_1|\psi_1\rangle_{\mathcal{H}_1}|^2 \, |\langle a_2|\psi_2\rangle_{\mathcal{H}_2}|^2. \end{aligned} \tag{8.33}$$

On the other hand, the single system predictions are

$$\begin{aligned} \mathrm{Prob}(A_1 = a_1; \, |\psi_1\rangle) &= |\langle a_1|\psi_1\rangle_{\mathcal{H}_1}|^2 \\ \mathrm{Prob}(A_2 = a_2; \, |\psi_2\rangle) &= |\langle a_2|\psi_2\rangle_{\mathcal{H}_2}|^2 \end{aligned} \tag{8.34}$$

and hence

$$\mathrm{Prob}(A_1 = a_1, A_2 = a_2; \, |\psi_1\rangle \, |\psi_2\rangle) = \tag{8.35}$$

$$\mathrm{Prob}(A_1 = a_1; \, |\psi_1\rangle) \, \mathrm{Prob}(A_2 = a_2; \, |\psi_2\rangle). \tag{8.36}$$

This result is the physical justification for choosing Eq. (8.28) as the definition of the scalar product on the tensor product space.

4. The tensor product is used to construct the full Hilbert space of a spinning electron. The spatial modes (taking one-dimensional motion for

simplicity) of this particle are described in the usual way by wave functions $\psi \in L^2(\mathbb{R})$, and the spin modes use the Hilbert space $\mathbb{C}^2$. The Hilbert space that describes both modes is the tensor product $L^2(\mathbb{R}) \otimes \mathbb{C}^2$, *i.e.*, the spatial and spin modes are treated as if they were subsystems of the full electron system. An important set of product vectors is the set of the type $|x\rangle|\uparrow\rangle$ or $|x\rangle|\downarrow\rangle$, where $|x\rangle$ denotes the (generalised) eigenstate of $\widehat{x}$ with continuous eigenvalue x. Thus, for example, $|x\rangle|\uparrow\rangle$ corresponds to an electron whose position is x, and whose z component of internal spin is $+\frac{1}{2}\hbar$; and analogously for $|x\rangle|\downarrow\rangle$.

More rigorously, the general state of $L^2(\mathbb{R}) \otimes \mathbb{C}^2$ can be written in the form

$$\vec{\chi} := \psi_1 \otimes \begin{pmatrix} 1 \\ 0 \end{pmatrix} + \psi_2 \otimes \begin{pmatrix} 0 \\ 1 \end{pmatrix} \tag{8.37}$$

where $\psi_1, \psi_2 \in L^2(\mathbb{R})$, and where $\binom{1}{0}$ and $\binom{0}{1}$ are the eigenstates $|\uparrow\rangle$ and $|\downarrow\rangle$ in the basis for $\mathbb{C}^2$ in which the operator representing S_z is $\widehat{S}_z = \frac{1}{2}\hbar \begin{pmatrix} 1 & 0 \\ 0 & -1 \end{pmatrix}$.

8.4.2 The Idea of Quantum Entanglement

Quantum entanglement is associated mathematically with the existence of vectors in $\mathcal{H}_1 \otimes \mathcal{H}_2$ that are *not* of the simple product form $\vec{\psi}_1 \otimes \vec{\psi}_2$, and which therefore cannot be associated with vector states of (suitable ensembles of) the subsystems S_1 and S_2. Of course, one possibility is that states of this type never arise in practice: a possibility that we shall refute by considering the role of quantum entanglement (and its relation to measurement theory) in the example of the measurement[9] of the spin of an electron along the z-axis.

The result of such a measurement will be either 'spin-up' or 'spin-down', corresponding to $S_z = \frac{1}{2}\hbar$ and $S_z = -\frac{1}{2}\hbar$ respectively, with eigenvectors denoted $|\uparrow\rangle$ and $|\downarrow\rangle$. Now suppose that the apparatus itself can be described in quantum-mechanical terms: a supposition that seems not

[9]The phenomenon of quantum entanglement is not peculiar to a measurement interaction: it arises in many other situations as well. However, the example of measurement is particularly graphic, and it leads directly to the infamous 'measurement problem' in quantum theory.

unreasonable since all matter is built from atoms. Of course, if the apparatus is to function as a proper measuring device, its state must change in an appropriate way when it couples to the electron.

To see how this works, suppose there is some apparatus observable (for example, the position of a pointer on a dial) that has three possible values A_0, $A_\uparrow$ and $A_\downarrow$, where A_0 is the neutral, 'resting' position, and $A_\uparrow$ and $A_\downarrow$ are the values of the observable following a measurement of the electron having spin-up and spin-down respectively. We suppose that, in the quantum-mechanical description of the apparatus, there is some self-adjoint operator which has the three eigenvalues A_0, $A_\uparrow$ and $A_\downarrow$ with associated eigenvectors $|A_0\rangle$, $|A_\uparrow\rangle$ and $|A_\downarrow\rangle$.

Now consider a collection of many copies of the combined system of electron plus apparatus that is prepared to be in the quantum state $|A_0\rangle|\uparrow\rangle$. Thus, with probability one, a measurement by a second device of the apparatus pointer will give the resting value A_0, and similarly a measurement of the electron spin by a third device will give spin-up. In this probability-one situation, it is meaningful to say of each individual composite system that the pointer *has* value A_0, and the electron *has* spin-up.

During the interaction between apparatus and electron this state vector evolves according to the corresponding time-dependent Schrödinger equation in such a way that, when the measurement is finished, it has undergone the transformation

$$|A_0\rangle|\uparrow\rangle \longrightarrow |A_\uparrow\rangle|\uparrow\rangle \tag{8.38}$$

where we have assumed an interaction in which the initial configuration of the electron is not changed. Similarly, if the initial state is spin-down, the measurement operation can be described by the transformation

$$|A_0\rangle|\downarrow\rangle \longrightarrow |A_\downarrow\rangle|\downarrow\rangle. \tag{8.39}$$

The interpretation of these equations is straightforward. For example, the right hand side of Eq. (8.38) clearly describes a situation in which the apparatus has passed into the state corresponding to the registration of the fact that the spin was up, and the electron spin itself is as it was before the measurement was made; similar remarks hold for Eq. (8.39). Note that these remarks apply to the individual composite systems in the ensemble since the right hand sides of Eq. (8.38–8.39) are eigenstates of the appropriate operators.

But suppose now that the initial state of the electron is *not* an eigenstate of $\widehat{S}_z$ but is instead some linear superposition $\alpha|\uparrow\rangle + \beta|\downarrow\rangle$ (normalised so that $|\alpha|^2 + |\beta|^2 = 1$). It is easy to produce states of this sort: for example, $|\uparrow\rangle + |\downarrow\rangle$ is an eigenvector of $\widehat{S}_x$ pointing along the x-direction. Thus all we need to do to prepare $|\uparrow\rangle + |\downarrow\rangle$ is to precede the S_z-measurement with a state preparation for S_x, and then select the systems associated with the result $\frac{1}{2}\hbar$. Note that this technique works in general since any vector of the form $\alpha|\uparrow\rangle + \beta|\downarrow\rangle$ is the eigenvector of the component of the spin along *some* axis.

The quantum state of the ensemble of composite systems before the measurement is $|A_0\rangle(\alpha|\uparrow\rangle + \beta|\downarrow\rangle)$ and, since the Schrödinger time evolution is *linear*, this must evolve under the measurement interaction into the appropriate *sum* of the results in Eq. (8.38) and Eq. (8.39); that is

$$\boxed{|A_0\rangle(\alpha|\uparrow\rangle + \beta|\downarrow\rangle) \longrightarrow \alpha\,|A_\uparrow\rangle|\uparrow\rangle + \beta\,|A_\downarrow\rangle|\downarrow\rangle.} \tag{8.40}$$

A concrete example of this situation is afforded by the Stern–Gerlach device, in which an incoming spin-half atom is moved either along, or opposite to, the direction of the gradient of the magnetic field according to the value of the component of the spin along this axis. It is feasible to regard the final position of the atom as an analogue of the apparatus pointer above; *i.e.*, if the two possible positions of the atom when it hits the screen (or whatever) are $z_\uparrow$ and $z_\downarrow$, then knowing this position is equivalent to knowing the spin value.[10] But then, if the spin state of the incoming atom as it enters the Stern–Gerlach apparatus is some linear *superposition* $|z_0\rangle \otimes (\alpha|\uparrow\rangle + \beta|\downarrow\rangle)$, the outgoing atom will be described by the linear combination $\alpha\,|z_\uparrow\rangle|\uparrow\rangle + \beta\,|z_\downarrow\rangle|\downarrow\rangle$ just before it hits the screen.

The crucial conceptual question is the meaning of the right hand side of Eq. (8.40). The final state is a *superposition* of product vectors and,

[10]This remark is a little oblique. As mentioned earlier, it can be argued that a Stern–Gerlach device performs a state preparation that is *not* a measurement, ideal or otherwise. Thus it might be more accurate to say that a subsequent *measurement* of the atom position will reveal the spin value. In this sense, the Stern–Gerlach device is not a good example of a measuring system being treated quantum mechanically. However, it does show clearly how an entangled state can arise.

as such, does not correspond to either the atom or the apparatus (or any ensemble of such) being associated with any specific state. In particular, no state reduction has taken place, and both possible outcomes of the measurement process have an equal status. This is a good example of 'quantum entanglement': even if a composite system starts with each of two subsystems being meaningfully associated with a state (*i.e.,* the initial composite state is a product vector), after they have interacted the state of the composite system will typically have become a linear superposition of product states. As we shall see in chapter 9, this phenomenon of quantum entanglement plays a fundamental role in the formulation of the famous Einstein, Podolsky and Rosen paradox.

The phenomenon of entanglement might be acceptable if both subsystems are atomic size or smaller; indeed, according to the Pauli exclusion principle, any pair of electrons necessarily exists in a certain type of entangled state. However, this picture seems very peculiar when, as above, one of the subsystems is a macroscopic piece of equipment. The infamous Schrödinger-cat[11] paradox is another example of this type.

Clearly this problem is related to the question of whether or not a macroscopic system can exist in a linear superposition of eigenstates of the collective observables that describe its macroscopic properties and, if it can, what this means. However, the problem of quantum entanglement is even worse in the sense that a macroscopic subsystem (or ensemble of such) may not obviously be described by any state at all, not even a linear superposition of eigenstates!

The phenomenon of quantum entanglement suggests a holistic structure for the physical world that contrasts strongly with the predominantly reductionist views of Western philosophy whereby composite systems may be analysed in terms of their constituent subsystems. It also seems at variance with our actual experience of how large objects behave: a behaviour that is arguably one of the main reasons why Western philosophy has evolved the way it has. As a consequence, there have been a number of attempts to change the mathematical structure of quantum theory in such a way that, for example, a macroscopic object automatically disentangles itself from anything with which it has interacted in the past. Or, to put it another way, some sort of objective reduction of the state vector takes

[11]A translation of Schrödinger's famous paper can be found in Wheeler & Zurek (1983).

place for such systems as part of their dynamical evolution. This, and other, attempts to resolve the measurement problem will be discussed in the next section.

8.5 The Measurement Problem

8.5.1 The Nature of the Problem

At its most general level, the 'measurement problem'[12] is to understand all aspects of the role of measurement in the foundations of quantum theory. Needless to say, the precise way in which this challenge is construed depends strongly on the overall interpretation of the formalism. In practice, the most studied issue is how the entangled, Schrödinger time evolution of object plus apparatus can be reconciled with the picture of a reduction of the state vector. This particular question takes on different forms according to whether the reduction concerned is of the type Eq. (8.1), Eq. (8.4), or Eq. (8.7)—*i.e.*, with selection of the results of the measurements (in the language of the instrumentalist)—or of the type Eq. (8.3) or Eq. (8.8)—where no selection is made.

The reduction Eq. (8.1), $|\psi\rangle \mapsto |a_m\rangle$, poses no problems for the pragmatist or instrumentalist since it is associated with a selective state preparation: an operation that is clearly grounded outside the system itself, and for which there is no reason to seek any relation with a Schrödinger time-evolution. True, even the most devoted instrumentalist may sometimes feel tempted to look for a scientific answer to the question of why the *actual* result obtained in any single measurement is what it is, and this might perhaps be construed as one aspect of the 'measurement problem'. However, such a quest would go well beyond the declared scope of these particular interpretations of quantum theory.

The problem posed to a *realist* by $|\psi\rangle \mapsto |a_m\rangle$ is quite different. Indeed, one of the major challenges for any such interpretation of quantum theory is to explain how it is that anything *happens* at all: where do 'actual facts' come from? In particular, the realist may want to view the

[12] A comprehensive source of original papers and comments on the measurement problem is Wheeler & Zurek (1983). For a modern analysis see Busch et al. (1991).

reduction $|\psi\rangle \mapsto |a_m\rangle$ as (i) an objective change in the state of an individual system; and (ii) something that is directly associated with the 'coming-to-be' of the fact "A has the value a_m". In an instrumentalist approach to quantum theory, the transition from potentiality to actuality is associated with a measuring device (ideal or otherwise) whose own internal description necessarily lies *outside* the domain of the theory (or, at least, does so in so far as it fulfills the measurement role assigned by the instrumentalist philosophy). But this path is closed in any interpretation of quantum theory that aspires to regard measurement as just a particular type of physical interaction, and therefore something that should be fully describable in terms of the formalism itself.

The issues surrounding the second type of reduction, in which $|\psi\rangle \mapsto \rho_{\text{red}} = \sum_{m=1}^{M} |\langle a_m|\psi\rangle|^2 \widehat{P}_{|a_m\rangle}$, are rather different from those relating to $|\psi\rangle \mapsto |a_m\rangle$, and are usefully illustrated by the example of the measurement of electron spin. As shown above, viewing the apparatus plus electron as a composite quantum system that evolves according to the Schrödinger equation gives the entangled vector state

$$|\psi_{\text{entan}}\rangle := \alpha\,|A_\uparrow\rangle|\uparrow\rangle + \beta\,|A_\downarrow\rangle|\downarrow\rangle, \tag{8.41}$$

whereas using the idea of making ideal spin measurements with no selection of results gives rise to the density matrix

$$\rho_{\text{red}} = |\alpha|^2\,|\uparrow\rangle\langle\uparrow| + |\beta|^2\,|\downarrow\rangle\langle\downarrow|. \tag{8.42}$$

In this case, the problems circle around a feeling that these two, quite different, states must be compatible in some way since both relate to a situation in which no *particular* measurement result for the electron has been recorded. However, they are certainly not the same: Eq. (8.41) is a *vector* state in the Hilbert space of the composite system, whereas Eq. (8.42) is a *mixed* state in the Hilbert space of the electron alone.

That these states differ causes no surprise to an advocate of the pragmatic approach since they are derived therein in two completely different ways. The only problem is to show that they are mutually consistent within the self-limited framework of this approach. We shall discuss this issue shortly.

However, the challenge for a realist is different. The first problem is to reconcile the entangled state in Eq. (8.41) with the prejudice that a classical object like a measuring device must always *possess* values for its 'macroscopic' physical quantities. Note that this is not a problem for a pragmatist who will interpret Eq. (8.41) only in relation to the results of measurements made on the composite system; see below.

Secondly, a realist who assigns states to individual systems will say that, (i) if it occurs at all, the reduction to Eq. (8.42) (or, perhaps, to something to which Eq. (8.42) is an approximation) applies to a single copy of apparatus-plus-electron; and (ii) the reduction must arise from an internal dynamical change in this closed system. Thus the challenge is to see if Eq. (8.42) can be *rederived* in some way from the entangled vector Eq. (8.41) that results from the object plus apparatus being treated as a composite quantum system.

This particular problem is compounded by the fact that the time-dependent Schrödinger equation can never lead to a vector state $|\psi\rangle$ (or, equivalently, a density matrix of the special type $\widehat{P}_{|\psi\rangle}$) evolving into a non-trivial density matrix (*i.e.*, one that is not of the form $\widehat{P}_{|\phi\rangle}$ for some vector $|\phi\rangle$) [Exercise]. Thus it is difficult to see how a transition like Eq. (8.3) could ever be described exactly using the standard quantum formalism: the best that can be hoped for is that an entangled state like Eq. (8.41) will be equivalent to the mixed state Eq. (8.42) "for all practical purposes".

8.5.2 The Pragmatic View

As remarked above, from the perspective of the pragmatic approach there is no real measurement problem. The states Eq. (8.41) and Eq. (8.42) are obtained in two completely different ways—the first from an internal Schrödinger time evolution; the second from a series of repeated ideal measurements whose results are ignored after the selection process—and there is no particular reason why these should be equal.

On the other hand, there is a need to demonstrate a certain *consistency* between the two states Eq. (8.41) and Eq. (8.42). According to the relative-frequency interpretation, the full physical significance of the entangled state Eq. (8.41) can only be extracted by taking a large collection of copies of apparatus-plus-system on which a series of repeated measurements are made with a *second* device that is external to both. In particular, if a series of repeated (ideal) measurements is made of, for example, the apparatus, and if a selection is made of those copies of the composite system in which the result $A_\uparrow$ is found (predicted by Eq. (8.41) to be a fraction $|\alpha|^2$ of the total), then the state appropriate for predicting the results of further measurements on this subset of systems is the reduced vector $|A_\uparrow\rangle|\uparrow\rangle$. Therefore, if a measurement of the z-component of the electron spin is made on each element of this subset, the theory predicts that the result will necessarily be $+\frac{1}{2}\hbar$. Similarly, the theory predicts a 100% correlation between finding the result $A_\downarrow$ for the apparatus, and $-\frac{1}{2}\hbar$ for the electron spin. Thus the predictions for the spin results are consistent with Eq. (8.42).

This process can be extended indefinitely with a measurement by a third device of the second device-plus-apparatus-plus-system (assuming that the second device performed an ideal measurement), and then a fourth device measuring the third device-plus-... and so on. Thus the minimal, pragmatic approach is *consistent* in its own terms with the idea that it is possible to treat real measuring devices in quantum-mechanical terms, but it cannot go beyond this point.

8.5.3 Some Conventional Resolutions

One approach to the measurement problem for a realist is to claim that, in fact, there is no such problem: on the contrary, the entangled state Eq. (8.41) is precisely what is *needed* to 'explain' the mixed state Eq. (8.42). The argument (which, technically, is another version of the consistency result discussed above) runs as follows. Suppose we decide to study a physical quantity B for the electron; how do the predictions given by the two states differ? In the Hilbert space of the composite system, B is represented by the operator $\hat{\mathbb{1}}\otimes\hat{B}$, and hence

$$\langle B\rangle_{\psi_{\text{entan}}} = \langle\psi_{\text{entan}}|\,\hat{\mathbb{1}}\otimes\hat{B}\,|\psi_{\text{entan}}\rangle = |\alpha|^2\langle\uparrow|\hat{B}|\uparrow\rangle + |\beta|^2\langle\downarrow|\hat{B}|\downarrow\rangle, \quad (8.43)$$

where we have used the orthogonality relation $\langle A_\uparrow|A_\downarrow\rangle = 0$. On the other hand, the prediction for the expected value of B when the state is the reduced density matrix Eq. (8.42) is

$$\begin{aligned}\langle B\rangle_{\rho_{\text{red}}} &= \operatorname{tr}(\hat{\rho}_{\text{red}}\hat{B}) = |\alpha|^2\operatorname{tr}(|\uparrow\rangle\langle\uparrow|\hat{B}) + |\beta|^2\operatorname{tr}(|\downarrow\rangle\langle\uparrow|\hat{B}) \\ &= |\alpha|^2\langle\uparrow|\hat{B}|\uparrow\rangle + |\beta|^2\langle\downarrow|\hat{B}|\downarrow\rangle, \quad (8.44)\end{aligned}$$

which is the same as Eq. (8.43).

This result illustrates a certain consistency between the two states (as did the analogous argument for a pragmatist or instrumentalist) but it is fallacious to say that they are completely equivalent. The equality of the probabilistic predictions of the entangled state and reduced density matrix holds only for physical quantities that pertain to the electron alone; *i.e.*, special operators of the form $\widehat{\mathbb{1}} \otimes \widehat{B}$. Physical quantities involving both electron and apparatus will reveal differences between these two states. Consequently, the mixed state that is effectively obtained in Eq. (8.43) by 'tracing out' the degrees of freedom of the apparatus *cannot* be given any simple interpretation in terms of possessed, but unknown, values for the electron spin; for this reason, mixed states of this type are sometimes called 'improper' mixtures. Thus, if the strict instrumentalist view is felt to be insufficient, the problem of reconciling entangled state and reduced density matrix remains. In particular, this is so for any interpretive scheme that aspires to associate quantum states with individual systems.

Some of the possible responses to this situation are as follows.

1. The mathematics of standard quantum theory is not universally applicable. The use of state vectors and self-adjoint operators is correct, but the 'real' theory involves a non-linear time evolution in which unacceptable states never arise. In particular, entangled states never occur for macroscopic systems, or in any other situation where their presence would embarrass our philosophical prejudices. This includes real measuring devices, and cats (be they dead or alive). Linear superpositions of eigenstates of macroscopic properties also never arise in such systems.

2. The mathematics of standard quantum theory is universally applicable, and hence entangled states like Eq. (8.40) do arise. However, these can be interpreted in such a way that, for all practical purposes, it is *as if* a state reduction has taken place.

3. Present day quantum theory has no fundamental status at all. It is at best an instrumentalist tool for predicting the results of certain types of external operation made on certain types of system, but it cannot cope with the sorts of problem involving closed systems we have been discussing here. Something radically new is needed: far

more than just changing the time-evolution equation to one that is non-linear.

The first approach essentially posits the existence of special interactions that cannot be described using the standard quantum formalism, and which lead to a genuine disentanglement of the state vector as a result of a dynamical process in some new, extended theory. Thus, in terms of each individual system, it is this special interaction that produces the transition from potentiality—as described by an entangled state vector like Eq. (8.40)—to the reality in which one *actual* outcome is manifest. Note that there is no logical reason why this reductive step should be deterministic: the new 'dynamical process' could be a stochastic process of some sort, in which case the final state would be a density matrix of the type we have been considering.

A number of concrete suggestions have been made about such mechanisms, although opinions differ widely on the origin of the non-linearity, or the scale at which it acts. Some of the more plausible suggestions are as follows.

- Quantum theory becomes non-linear when applied to 'large' systems of macroscopic or near-macroscopic size (for example: pieces of laboratory equipment, vice-chancellors, or cats). Several theories of this type have been proposed that assume some stochastic process (rather like Brownian motion) which generates the reduction. Relatively recent examples that have caused much interest are the work of Ghirardi, Rimini & Weber (1986), and of Pearle (1986).

- The source of non-linearity is gravity considered in the context of general relativity. Once again there is an implication that the size of the system is a key factor in determining the degree to which the evolution is non-linear. A particular idea of this type is discussed in the fascinating book by Penrose (1989). Penrose's central idea is that quantum theory breaks down at the point at which the conventional theory would produce something like a superposition of eigenstates of space-time geometry in a quantum theory of gravity.

- The process of describing measurements by standard quantum theory works up to the point at which the results are acquired by a

> human mind, but there the formalism breaks down. It was von Neumann who first made this startling suggestion that the human mind is the ultimate measuring device: *i.e.*, the movement from potential to actual finally takes place when the result of a measurement interaction reaches human consciousness. However, he felt that the functioning of the mind itself is intrinsically beyond the reach of theoretical physics. Of course, such a position leaves completely open a number of very difficult questions about the meaning of 'consciousness', the problem of mind-body duality, and the like.
>
> A different view was taken by Wigner (1961) who asserted that, in principle, the functioning of a mind *can* be described in physical terms, but this requires a *non*-linear version of quantum theory so that the human mind never 'finds itself' in a superposition of quantum states. Presumably a cat (at least, one that is alive) is also sufficiently conscious to enjoy this, no-doubt desirable, property.
>
> Wigner later changed his mind on this issue, but the general approach has been a popular one and, in various forms, has been advocated frequently over the years; indeed, sometimes it has even been referred to as the 'standard' interpretation of quantum theory. More recent advocates of the use of consciousness include Stapp (1993), and Lockwood (1992) (the latter mainly in the context of the many-worlds interpretation; see below).

The invocation of consciousness as a fundamental category raises profound issues that go well beyond the normal remit of physics, and it is desirable therefore to look for alternative resolutions of the measurement problem. In addition, there are some situations where it is very difficult to see how the role of consciousness can usefully be applied at all. An obvious example is quantum cosmology, where one would like to talk of quantum states of the entire universe, and without the need to invoke the presence of a 'universal mind' that 'makes things happen'.

The desire to find alternative resolutions of the measurement problem has resulted in much attention being devoted in recent years to the second type of approach in which the present technical formulation of quantum theory (including Eq. (8.40)) is assumed to be correct, but the picture of a state reduction is now viewed as a heuristic device that applies in an approximate way only. In particular, it is claimed that the coupling of a system to its environment can lead to an apparent reduction of the state vector for the system, even though all dynamical evolution is described by a linear Schrödinger equation. This process is especially effective for macroscopic systems, and means that linear superpositions of eigenstates of macroscopic observables never arise in practice; or, more precisely, if they do, they disappear very rapidly.

In many respects, this proposal looks very promising. The effective reduction takes place by a process of *decoherence*, which employs the mathematical operation (mentioned earlier in the context of Eq. (8.43–8.44)) whereby a vector state for a pair of systems becomes a mixed state for one of them if the degrees of freedom of the other are 'traced over', *i.e.*, if we feed in our ignorance of the detailed configurations of a complex subsystem. Of particular relevance here is the close-packed nature of the eigenvalues of a typical observable in a large system, which results in almost any coupling to its environment producing significant changes. In effect, the notion of an 'isolated system' (which is posited in formulating the rules of quantum theory) is an idealisation that can never really apply: the only true isolated system is the universe itself.

A number of studies have shown how very quickly an effective reduction of this sort takes place for a macroscopic system: even the interaction with the 3^0-Kelvin microwave radiation is sufficient to produce decoherence in around 10^{-23} seconds (Joos & Zeh 1985)! However, such a resolution of the problem of entangled states is subject to the same objection as that mentioned in the context of Eq. (8.43–8.44): namely, the density matrix thus produced does not lead to a *bona fide* ignorance interpretation of possessed properties of the macroscopic system.

This objection might be sidestepped in the context of normal physics, but it certainly comes into play if one insists on including *all* the accessible information about the composite system, however difficult this might be in practice. The only really acceptable position would be one in which it could be shown that, even in *principle*, the lost correlations between environment and system can never be recovered. For example, one might invoke considerations peculiar to certain types of cosmology that arise in the context of general relativity.

Certainly the most acute form of the reduction problem arises in quantum cosmology, where attempts are made to describe the quantum states of the entire universe, including the gravitational field which, in all other branches of physics, forms part of the fixed space-time background. Of course, in this situation there is no environment at all. In truth, the heart of the matter is that we have still not succeeded in understanding how the transition from 'potential' to 'actual' really comes about.

8.5.4 Many Worlds

A more dramatic attempt to resolve the measurement problem is the *many-worlds* interpretation, which insists that *all* changes in time take place according to the linear Schrödinger evolution. Thus there are no state-vector reductions in any objective sense, and so, for example, equation Eq. (8.40) is correct as it stands. In addition, all physical systems are to be described in quantum-mechanical terms, and this includes what we normally regard as a measurement interaction. Thus 'measurement' as such plays no fundamental role in the theory; in particular, there is no *a priori* existing classical realm within which the quantum theory is to be interpreted. Instead, the usual ideas of measurement and classical physics have to 'emerge' from the quantum formalism in some way.

This approach to quantum theory dates back[13] to an important paper by Everett (1957) and has become increasingly popular in recent years, especially amongst quantum cosmologists who, for obvious reasons, welcome the absence of any fundamental structural role for acts of measurement. Indeed, it is rather difficult to think of *any* interpretation of quantum cosmology that does not invoke this view in one way or another. Thus 'post-Everett' schemes have become almost obligatory for those working in the physics of the very early universe.

The key feature of the many-worlds interpretation is the meaning given to an entangled state like Eq. (8.41). Specifically, it is claimed that both terms in Eq. (8.41) describe something that really exists: the state does *not* just refer to the probabilities of results that would be obtained if some external measurements were made on the composite system. However, different authors have interpreted the phrase "really exists" in different ways. DeWitt famously developed the idea that the different terms in an

[13]Curiously enough, the 'many worlds' aspect seems to have been anticipated in a novel by Charles Williams (1968), first published in 1931. Charles Williams, who was a close friend of J.R.R. Tolkien and C.S. Lewis, wrote several rather strange and moving books. They are recommended reading for would-be gnostics.

entangled state can be interpreted as showing that the universe *branches* into a number of different worlds (DeWitt & Graham 1973). Thus, for example, if applied to Eq. (8.41), the picture is of a branching into two worlds during the course of the measurement interaction: in one of these spin-up has occurred; in the other, spin-down.

On the other hand, Everett himself placed the emphasis on the role of subjective states for observers that were correlated with various aspects of the rest of the material universe. From this perspective, the theory deals more with many 'viewpoints', rather than many 'worlds'. One recent development of this particular perspective is the so-called the 'many minds' interpretation of quantum theory (for example, Lockwood (1992)).

One obvious response to such claims is to ask why we do not feel ourselves continuously branching into these different universes (or, in the many viewpoints interpretation, why we only experience one single view). In replying to this challenge, a supporter of the many-worlds interpretation will point to the internal consistency of quantum theory when applied to collections of coupled systems—a consistency on which we have commented already in the context of measuring both system and apparatus with a third device when the state of the composite system is an entangled vector like Eq. (8.41). In particular, reference is often made to *memory* states, which are meant to model the processes by which we (or computers) store information. Thus the main claim is that we do not feel ourselves splitting because, in every branch of the state-vector, there is a 100% correlation between our memory states and other events that have 'occurred'.[14]

The central point is that what is really important are the *correlations* between various properties of a system. Thus, for example, it would be claimed that the real meaning of the state vector in Eq. (8.41) is that, if the electron 'has' spin-up, then necessarily the apparatus will be in the state $|A_\uparrow\rangle$, and similarly for spin-down. Of course, this prediction would also be made in conventional quantum theory, but there it appears as a *conditional* probability on the results of measurements made on apparatus-plus-electron. The exact sense in which the new picture can be said to involve 'many' worlds (or worlds that 'branch') depends on the precise

---

[14]Everett's approach contains several implicit assumptions about the nature and status of an 'observer', particularly in regard to the idea of memory; see Bohm & Hiley (1993) for a clear discussion of this and related matters.

meaning given to the word 'has' when one says that an electron 'has' spin-up.

The principle technical ingredient in the many-worlds interpretation is the idea of a *relative* state. Thus, if a state of a compound system of two subsystems is written in the form

$$|\psi\rangle = \sum_{i=1}^{N_1}\sum_{j=1}^{N_2} \psi_{ij}\,|\chi_i\rangle\,|\phi_j\rangle \tag{8.45}$$

then $\sum_{j=1}^{N_2} \psi_{ij}\,|\phi_j\rangle$ is said to be the state of system 2 *relative* to the state $|\chi_i\rangle$ of system 1. In the example of Eq. (8.41), the state of the electron relative to the state of the apparatus $|A_\uparrow\rangle$ or $|A_\downarrow\rangle$ is $|\uparrow\rangle$ or $|\downarrow\rangle$ respectively. In general, the index $i = 1, 2, \ldots, N_1$ labels the different branches that are defined in relation to the collection of states $|\chi_i\rangle$, $i = 1, 2 \ldots, N_1$, of system 1.

However, an obvious problem with this idea is that the decomposition Eq. (8.45) is not unique. A simple example (which will be used later in our discussion of the Einstein, Podolsky and Rosen paradox) is given by a pair of electrons whose total angular momentum is zero. The spin part of such a state can be written in the form

$$|\psi\rangle = \frac{1}{\sqrt{2}}(|\uparrow\rangle|\downarrow\rangle - |\downarrow\rangle|\uparrow\rangle) \tag{8.46}$$

where $|\uparrow\rangle|\downarrow\rangle$ is the state in which particles 1 and 2 have spin $+\frac{1}{2}\hbar$ and $-\frac{1}{2}\hbar$ respectively along the z axis, and *vice versa* for $|\downarrow\rangle|\uparrow\rangle$; the entangled nature of the state Eq. (8.46) reflects the Pauli exclusion principle for fermions. From the perspective of the many-worlds interpretation, the production of such a state describes a 'splitting' into two branches: one in which the first particle has $S_z = \frac{1}{2}\hbar$ and the second has $S_z = -\frac{1}{2}\hbar$; and one in which these spin assignments are reversed.

But consider now what happens if we focus instead on the x component of spin. The spin operators are

$$\widehat{S}_z = \frac{\hbar}{2}\begin{pmatrix} 1 & 0 \\ 0 & -1 \end{pmatrix}, \qquad \widehat{S}_x = \frac{\hbar}{2}\begin{pmatrix} 0 & 1 \\ 1 & 0 \end{pmatrix} \tag{8.47}$$

and the eigenvectors of $\widehat{S}_z$ are

$$|\uparrow\rangle = \begin{pmatrix} 1 \\ 0 \end{pmatrix}, \qquad |\downarrow\rangle = \begin{pmatrix} 0 \\ 1 \end{pmatrix} \tag{8.48}$$

whilst those of $\widehat{S}_x$ are

$$|\rightarrow\rangle = \frac{1}{\sqrt{2}}\begin{pmatrix}1\\1\end{pmatrix}, \qquad |\leftarrow\rangle = \frac{1}{\sqrt{2}}\begin{pmatrix}1\\-1\end{pmatrix} \tag{8.49}$$

where $|\rightarrow\rangle$ and $|\leftarrow\rangle$ correspond to eigenvalues $+\frac{1}{2}\hbar$ and $-\frac{1}{2}\hbar$, respectively. It is straightforward to show from Eq. (8.48–8.49) that the state $|\psi\rangle$ in Eq. (8.46) can be written in terms of the eigenvectors of $\widehat{S}_x$ as

$$|\psi\rangle = \frac{1}{\sqrt{2}}(|\leftarrow\rangle|\rightarrow\rangle - |\rightarrow\rangle|\leftarrow\rangle). \tag{8.50}$$

Applied to Eq. (8.50), the many-worlds interpretation implies that the 'split' is into branches in which it is the S_x observables whose values are correlated. On the other hand, the form Eq. (8.46) implies a splitting into branches in which the correlation is with respect to the values of S_z. Which, if either, of these statements is correct? It is hard to see how they can both be right since, although it is true that the right-hand sides of Eq. (8.46) and Eq. (8.50) are equal, this is certainly *not* the case for the individual terms: the product vector $|\uparrow\rangle|\downarrow\rangle$ is *not* equal to either of the product vectors $|\leftarrow\rangle|\rightarrow\rangle$ or $|\rightarrow\rangle|\leftarrow\rangle$.

In effect, we need some special basis set for the Hilbert space of one of the systems, with respect to which the branching occurs. Various attempts have been made to find such a basis: for example, the eigenvectors of the position of a macroscopic object like a pointer on a piece of equipment. The use of such a special basis might be justified pragmatically with the type of decoherence argument used earlier; *i.e.*, it is plausible to say that a pointer acquires an actual position very quickly as a result of decoherence with the environment. Other attempts have invoked cosmological considerations, sometimes associated with the idea that there may be preferred reference frames in physics that select special choices for time. Note that the question of 'time' is bound to arise in any discussion of branching because of the implication that the latter takes place at some particular time. This is particularly problematic in the context of quantum cosmology because of the infamous 'problem of time': the curious phenomenon whereby time tends to disappear completely in certain natural approaches to quantum gravity (for example, see Isham (1994)).

At a fundamental level the situation is still rather unclear, and perhaps the main lesson is that it is very misleading to describe the many-worlds interpretation of quantum theory in terms of the universe 'splitting' into different branches in some literal way. Unfortunately, the construction of a more meaningful description remains a matter of debate. For this reason, there has been much interest during the last decade in developing approaches to quantum theory that are post-Everett in the sense of affirming that the time-dependent Schrödinger equation always holds—*i.e.*, there are no external state 'reductions'—but which differ from standard quantum theory in other significant ways.

One of the most interesting ideas of this type is the *consistent histories* interpretation, which is based on a precise mathematical specification of when it is meaningful to ascribe a probability to a *history* of a system, rather than to the value of an observable at a fixed time. The basic philosophical position of this scheme is arguably 'many-worlds', but the technical framework is more sophisticated and developed than that attached to the earlier ideas. In particular, the awkward notion of 'branching' is replaced by a more coherent and mathematically precise picture.

Currently, this intriguing approach is generating much interest, not least because it offers new possibilities for handling the problem of time in quantum cosmology. However, it is beyond the scope of this book to develop the scheme here, and the interested reader is referred to a recent work by Omnès (1994), one of the originators of the programme.[15]

8.5.5 Hidden Variables?

Some of the suggestions above are distinctly bizarre and, as such, remain highly contentious and the subject of heated debate. It is not surprising therefore that, over the years, there has been considerable sympathy for the idea that the standard quantum formalism is incomplete, and that the probabilistic nature of the usual results arises from the existence of certain 'hidden variables'.

There are several different types of hidden variable theory, but a common characteristic is the assertion that, in any given quantum state $\vec{\psi}$, any observable A *possesses* an objectively existing value $A(\vec{\psi}, \lambda_1, \lambda_2, \ldots, \lambda_n)$ which is determined by $\vec{\psi}$ and the values of a set of hidden variables $\{\lambda_1, \lambda_2, \ldots, \lambda_n\}$ which belong to some space Λ. It is further assumed that

[15]For a review with an extensive bibliography see (Omnès 1992).

there exists some probability density $\mu_{\vec{\psi}}$ on Λ such that the expected value of A in the state $\vec{\psi}$ is

$$\langle A\rangle_{\vec{\psi}} = \int_\Lambda d^n\lambda\ \mu_{\vec{\psi}}(\lambda_1, \lambda_2, \ldots, \lambda_n)\, A(\vec{\psi}, \lambda_1, \lambda_2, \ldots, \lambda_n). \tag{8.51}$$

The central challenge is to find spaces Λ, probability densities $\mu_{\vec{\psi}}$, and value functions $A(\vec{\psi}, \lambda_1, \lambda_2, \ldots, \lambda_n)$ such that this expression reproduces the predictions of standard quantum theory.

Many of the conceptual problems mentioned above are ameliorated if hidden variables exist. For example, an objective reduction of the type Eq. (8.3) might be generated by a purely deterministic (but non-linear) evolution associated with a dynamics for the hidden variables. In general, hidden variables theories adopt a philosophical position that is similar to the realism of classical physics. However, hidden variables introduce problems of their own and result in a picture of physical reality which is arguably as peculiar as that of the conventional theory. This is a fascinating topic and leads naturally to the discussion, deferred until now, of precisely what goes wrong with any attempt to assign actual, *possessed*, values to all physical quantities. In particular, we must now become engaged with the Kochen–Specker theorem, quantum logic, the Einstein-Podolsky-Rosen paradox, and the famous Bell inequalities.

Chapter 9

PROPERTIES IN QUANTUM PHYSICS

9.1 The Kochen–Specker Theorem

Some, perhaps many, users of quantum theory may find the pragmatic approach (or even a full instrumentalist interpretation) satisfactory, but it is hard to believe that it gives the ultimate view of physical reality. Indeed, with its reluctance to grant objective status to properties of individual systems, it is arguable whether it gives a picture of reality at all. However, attempts to move towards a more realist philosophy by ascribing 'latent' properties to individual systems tend to be too vague to do much more than stimulate a search for new conceptual categories. In particular, there is the unresolved issue of how the transition from potentiality to actuality is made: a major issue in any area of quantum theory, and one that is of crucial significance in the case of quantum cosmology, where the individual system concerned is the whole universe.

It is not surprising, therefore, that many physicists have sought an interpretation of quantum theory in which the probabilistic results would have the same status as those in classical statistical physics where these difficult philosophical problems do not arise. Most discussions of this type involve the epistemic interpretation of probability, in which probabilistic assignments refer to our *knowledge* of an objectively existing state of affairs; for example, an expression like $\Delta_\psi A$ is a measure of our ignorance of A which, it is supposed, does have an actual value. Such a position is consistent with a thoroughly realist view of the world, in which individual objects and properties have an unequivocal existence that is independent of any act of observation or measurement. Interpretations of this type are naturally coupled with the assumptions that (i) quantum states refer to individual systems directly, not just to the outcome of repeated measurements, or to any associated ensemble of systems; and (ii) a (perfect) measurement of a physical quantity reveals the value that it possessed immediately beforehand.

Most of the problems discussed so far would not arise in a scheme where properties are possessed. For example, a state-vector reduction would merely reflect an acquisition of further knowledge about the system by making a measurement—a clear analogue of the classical idea of conditional probability. True, this raises once again the question of whether the measurement itself can be described in quantum-mechanical terms, but the significance of this problem is now no greater than it is in classical statistical physics.

Since an epistemic interpretation of quantum theory promises to remove most of the conceptual problems, one might wonder why the theory is not automatically presented in this way from the outset. The reason is that the concept of an individual system 'possessing' a value for all its physical quantities is difficult to reconcile with the actual formalism of quantum theory. The two main obstacles are the Kochen–Specker theorem, and the famous Bell inequalities.

Our first task is to discuss the Kochen–Specker theorem. This concerns the existence of a putative *value function* $V_{\vec{\psi}}(A)$ that is to be interpreted as the value of the physical quantity A when the quantum state (of an individual system) is $\vec{\psi}$. There is no difficulty with this concept in classical physics. As we saw in Section 4.2, to each physical quantity A there corresponds a function $f_A : \mathcal{S} \to \mathbb{R}$ (where $\mathcal{S}$ is the classical state space) such that the value of A in the state $s \in \mathcal{S}$ is just the value of f_A at s:

$$V_s(A) = f_A(s). \tag{9.1}$$

The situation in quantum physics is very different since we have no *prima facie* idea of how to specify the value of $V_{\vec{\psi}}(A)$ for any particular pair $(A, \vec{\psi})$. If $\vec{\psi}$ is an eigenvector $\vec{u}_a$ of the self-adjoint operator $\widehat{A}$ that represents A, then we might be inclined to say that $V_{\vec{u}_a}(A) = a$, where a is the corresponding eigenvalue, but it is unclear how to go beyond this special case. As a consequence, nothing useful can be said about the existence of quantum value-functions without postulating further properties for them.[1]

The most natural condition to impose is that, for any function $\mathcal{F} : \mathbb{R} \to \mathbb{R}$,

$$V_{\vec{\psi}}(\mathcal{F}(A)) = \mathcal{F}(V_{\vec{\psi}}(A)). \tag{9.2}$$

Thus, in any quantum state $\vec{\psi}$, the value of a function of a physical quantity is equal to the function evaluated on the value of the quantity; for example, the value of L_x^2 is the square of the value of the angular momentum L_x.

At this point we should recall the discussion in Section 5.2 about the meaning of the expression $\mathcal{F}(A)$. If we are 'quantising' a given classical system with state space $\mathcal{S}$ then $\mathcal{F}(A)$ can stand for $\mathcal{F}(f_A)$, where $f_A : \mathcal{S} \to \mathbb{R}$, and where the map $\mathcal{F}(f_A) : \mathcal{S} \to \mathbb{R}$ is defined by $\mathcal{F}(f_A)(s) := \mathcal{F}(f_A(s))$ for all $s \in \mathcal{S}$, as in Eq. (5.20). In this case, it is viable to regard a value function $V_{\vec{\psi}}$ as being defined on real-valued functions f on the classical state space $\mathcal{S}$.

However, if we do not wish to assume any such classical background then something different is needed. One possibility is to define $\mathcal{F}(A)$ as that physical quantity which is associated with the operator $\mathcal{F}(\widehat{A})$. Of course, this is only meaningful if the quantisation map from physical quantity to self-adjoint operator is (i) 'onto', so that $\mathcal{F}(\widehat{A})$ is the quantum representative of *some* physical quantity; and (ii) one-to-one, so that the quantity thus defined is *unique*. Under these circumstances, nothing is

---

[1]There are plenty of functions that satisfy the eigenvector requirement. For example, the function $V_{\vec{\psi}}(A) := \langle\vec{\psi}, \widehat{A}\vec{\psi}\rangle$ does so (assuming that $\vec{\psi}$ is normalised). However, as we know, this gives the *expected* value of A, so it is not a good choice for a value function in any situation in which the dispersion $\Delta_{\vec{\psi}}(A)$ is non-zero. Therefore, to get a sensible notion of a value function it is necessary to go beyond the eigenvector requirement.

lost by thinking of $V_{\vec{\psi}}$ as a function of self-adjoint operators, rather than physical quantities.[2]

Note that we are driven towards such a strategy because, on the one hand (and unlike in the pragmatic approach), an operational definition in terms of *measurements* is not appropriate in the context of the more realist interpretation being sought; but, on the other hand, we must avoid *defining* $\mathcal{F}(A)$ to be the physical quantity that satisfies Eq. (9.2) for all $\vec{\psi} \in \mathcal{H}$, since then Eq. (9.2) would just be a tautology. The potential trap is that this is precisely how a function of a physical quantity *is* defined in classical physics, as shown by Eq. (5.20). This discussion impinges also on the basic function-preserving requirement $\widehat{\mathcal{F}(A)} = \mathcal{F}(\widehat{A})$, which *does* become a tautology if $\mathcal{F}(A)$ is defined to be the physical quantity that is associated with the operator $\mathcal{F}(\widehat{A})$.

The intuitively plausible requirement Eq. (9.2) has the following important implications.

1. If $[\widehat{A}, \widehat{B}] = 0$, the value function is additive in the sense that, for all $\vec{\psi} \in \mathcal{H}$,

$$V_{\vec{\psi}}(A+B) = V_{\vec{\psi}}(A) + V_{\vec{\psi}}(B) \tag{9.3}$$

where $A+B$ denotes the physical quantity associated with the self-adjoint operator $\widehat{A} + \widehat{B}$.

To prove this, recall from Section 6.3 that $[\widehat{A}, \widehat{B}] = 0$ implies the existence of a self-adjoint operator $\widehat{C}$ such that $\widehat{A} = \mathcal{F}(\widehat{C})$ and $\widehat{B} = \mathcal{G}(\widehat{C})$ for some functions $\mathcal{F}, \mathcal{G} : \mathbb{R} \to \mathbb{R}$. Thus $\widehat{A} + \widehat{B} = (\mathcal{F}+\mathcal{G})(\widehat{C})$ and hence, according to the condition Eq. (9.2),

$$\begin{aligned} V_{\vec{\psi}}(A+B) &= V_{\vec{\psi}}((\mathcal{F}+\mathcal{G})(C)) = (\mathcal{F}+\mathcal{G})(V_{\vec{\psi}}(C)) \\ &= \mathcal{F}(V_{\vec{\psi}}(C)) + \mathcal{G}(V_{\vec{\psi}}(C)) = V_{\vec{\psi}}(\mathcal{F}(C)) + V_{\vec{\psi}}(\mathcal{G}(C)) \\ &= V_{\vec{\psi}}(A) + V_{\vec{\psi}}(B), \end{aligned} \tag{9.4}$$

[2]Another way of looking at this whole argument is to start with a value function $V'_{\vec{\psi}}$ that is a function of *self-adjoint operators*, rather than physical quantities. The analogue of the condition Eq. (9.2) is then the relation $V'_{\vec{\psi}}(\mathcal{F}(\widehat{A})) = \mathcal{F}(V'_{\vec{\psi}}(\widehat{A}))$, which deals only with mathematical entities. A value function $V_{\vec{\psi}}$ of physical quantities can then be defined by $V_{\vec{\psi}}(A) := V'_{\vec{\psi}}(\widehat{A})$. If the quantisation map is one-to-one and onto, then $\mathcal{F}(A)$ is well-defined as the inverse image of $\mathcal{F}(\widehat{A})$, and the $V_{\vec{\psi}}$ function constructed in this way satisfies Eq. (9.2).

as claimed.

2. Similarly, a value function possesses the multiplicative property that if $[\widehat{A}, \widehat{B}] = 0$ then, for all $\vec{\psi} \in \mathcal{H}$,
$$V_{\vec{\psi}}(AB) = V_{\vec{\psi}}(A)V_{\vec{\psi}}(B) \tag{9.5}$$
where AB denotes the physical quantity associated with the self-adjoint operator $\widehat{A}\widehat{B}$.

3. Let $\mathbb{1}$ denote the physical quantity that corresponds to the unit operator $\widehat{\mathbb{1}}$. Then, using Eq. (9.5) with $A := \mathbb{1}$, we see that, for each state $\vec{\psi}$, $V_{\vec{\psi}}(B) = V_{\vec{\psi}}(\mathbb{1})V_{\vec{\psi}}(B)$ for all physical quantities B. Thus, for each $\vec{\psi} \in \mathcal{H}$, we get
$$V_{\vec{\psi}}(\mathbb{1}) = 1 \tag{9.6}$$
provided only that there is at least one quantity B for which $V_{\vec{\psi}}(B) \neq 0$.

The multiplicative property Eq. (9.5) has an important implication for any physical quantity P whose representative is a projection operator $\widehat{P}$. Namely, the property $\widehat{P}^2 = \widehat{P}$ implies at once that $(V_{\vec{\psi}}(P))^2 = V_{\vec{\psi}}(P^2) = V_{\vec{\psi}}(P)$, and hence
$$V_{\vec{\psi}}(P) = 0 \text{ or } 1. \tag{9.7}$$
Thus, thinking of P as a *proposition*, we see that a quantum value function $V_{\vec{\psi}}$ gives it a 'false' or 'true' ascription in the state $\vec{\psi}$. We shall return to this feature in our discussion of quantum logic in Section 9.2.

Now consider a collection $\{\widehat{P}_1, \widehat{P}_2, \ldots, \widehat{P}_n\}$ of projection operators that forms a resolution of the identity; *i.e.*, $\widehat{P}_i\widehat{P}_j = 0$ if $i \neq j$, and
$$\widehat{P}_1 + \widehat{P}_2 + \cdots + \widehat{P}_n = \widehat{\mathbb{1}}. \tag{9.8}$$

For example, these could be the spectral projectors of a self-adjoint operator $\widehat{A}$ with discrete eigenvalues, so that $\widehat{P}_i = \widehat{P}_{A=a_i}$. In a realist interpretation, the associated proposition P_i asserts that A *has* a particular value a_i, which implies that the collection of propositions $\{P_1, P_2, \ldots, P_n\}$ should be mutually *exclusive*—only one can be true at any given time—and *exhaustive*—at least one of them must be true at any given time. This is in accord with the commonsense view of the nature of truth and falsity: in a set of mutually exclusive and exhaustive propositions one—and only one—can be true; the rest are false. This expectation is borne out by the formalism. Indeed, the results Eq. (9.3) and Eq. (9.6) imply that $\sum_{i=1}^{n} V_{\vec{\psi}}(P_i) = 1$ for all states $\vec{\psi}$. However, since each $V_{\vec{\psi}}(P_i)$ has the value 0 or 1, this sum can equal 1 only if one of the propositions is given the value 1, and the rest are given the value 0.

A special case is when $\widehat{P}_i := |e_i\rangle\langle e_i|$, where $\{|e_1\rangle, |e_2\rangle, \ldots\}$ is an orthonormal basis set for the Hilbert space $\mathcal{H}$. Then, according to the result just obtained, for any state $\vec{\psi}$ a value function $V_{\vec{\psi}}$ must assign the number 1 to one of the projectors/one-dimensional subspaces (more precisely, to the associated proposition) and 0 to the rest. This is not a trivial requirement, since any given vector will belong to many different orthonormal basis sets, and the value given to the corresponding one-dimensional subspace must be independent of the choice of such a set. In fact, this requirement is so difficult that it cannot be satisfied:

Theorem (Kochen and Specker) There is no such function $V_{\vec{\psi}}$ if the Hilbert space $\mathcal{H}$ is such that $\dim(\mathcal{H}) > 2$.

Proof

The ingenious proof of this theorem can be found in a number of texts. The original reference is Kochen & Specher (1967); good recent discussions of both the theorem and its implications can be found in Redhead (1989), Hughes (1989), and Peres (1993).

The first step is a straightforward demonstration that if a value function exists for a Hilbert space of a particular dimension N, then it necessarily does so for any space of dimension less than N. Similarly, if a value function exists for a complex Hilbert space, then one also exists on any real Hilbert space of the same dimension. Thus, to prove the theorem, it suffices to show that no such function can exist in real, 3-dimensional euclidean space.

Reducing the problem in this manner has the advantage that it can be studied in a manifestly geometrical way, in which case it looks like a certain type of map-colouring problem. In their original proof, Kochen and Specker found a counter example to the existence of a value function, in the form of a set of 117 vectors in the vector space $\mathbb{R}^3$ that could not be consistently assigned the numbers 0 or 1 in the desired way. More recently, Peres has produced an elegant example of 33 vectors (made up of 11 triads of orthogonal vectors in $\mathbb{R}^3$) which serve the same purpose.

The reader is referred to Peres (1991) and Peres (1993) for further details.

QED.

This remarkable result shows that a value function $V_{\vec{\psi}}$ cannot exist without violating one of the assumptions implicit in the statement of the theorem. The obvious candidates are:

1. the requirement Eq. (9.2) that $V_{\vec{\psi}}(\mathcal{F}(A)) = \mathcal{F}(V_{\vec{\psi}}(A))$;

2. the assumption that the quantisation map $A \mapsto \widehat{A}$ is one-to-one.

The natural situation in which the first requirement might, perhaps, be expected to fail is when we deal with a self-adjoint operator $\widehat{A}$ that can be written as functions $\widehat{A} = f_1(\widehat{A}_1)$ and $\widehat{A} = f_2(\widehat{A}_2)$ of a pair of self-adjoint operators $\widehat{A}_1$ and $\widehat{A}_2$ with $[\widehat{A}_1, \widehat{A}_2] \neq 0$. The value condition Eq. (9.2) would imply that

$$f_1(V_{\vec{\psi}}(A_1)) = f_2(V_{\vec{\psi}}(A_2)), \tag{9.9}$$

and, if $\widehat{A}_1$ and $\widehat{A}_2$ do not commute, the wisdom of this assumption could be questioned. For example, if we suppose that the value possessed by a quantity can be found by making an appropriate measurement (which *is* an assumption, albeit not an unreasonable one) then Eq. (9.9) implies that measuring A_1 and applying f_1 to the result, will yield the same number as measuring A_2 and applying f_2 to that result. However, there seems no good reason for supposing that this would be so if $[\widehat{A}_1, \widehat{A}_2] \neq 0$.

On the other hand, dropping the condition Eq. (9.9) has a very peculiar effect. For example, let $\widehat{A}_1$ and $\widehat{A}_2$ be a pair of non-commuting, self-adjoint operators whose spectral resolutions contain a common projector $\widehat{P}$. Then $\widehat{P}$ can be written as one function[3] of $\widehat{A}_1$, and as another function of $\widehat{A}_2$. Hence a failure of the value condition Eq. (9.2) (and hence of Eq. (9.9)) means that the value ascribed to the physical quantity P (or, equivalently, the truth or falsity of the associated proposition) depends on the *context* in which $\widehat{P}$ is taken: if viewed as belonging to the spectral representation of $\widehat{A}_1$, it will be given one value; if viewed as belonging to that of $\widehat{A}_2$, it will be given another.

[3]In general, if a self-adjoint operator $\widehat{A}$ has a discrete spectrum with spectral resolution $\widehat{A} = \sum_{i=1}^{M} a_i \widehat{P}_i$, then the projector $\widehat{P}_i$ can be written as the function of $\widehat{A}$, $\widehat{P}_i = \chi_{a_i}(\widehat{A})$ where $\chi_r(t) := 1$ if $t = r$ and 0 otherwise. The function χ appeared earlier in Eq. (5.31) for essentially this reason.

Equivalently, we can say that the value of P depends on what *other* physical quantities are assigned values at the same time; *i.e.*, it depends on a choice that is made for operators that commute with $\widehat{P}$. For example, if $\widehat{P}$ is thought of as belonging to the spectral representation of $\widehat{A}_1$, the remaining projectors in the spectral set are a natural choice for commuting partners; and analogously if $\widehat{P}$ is viewed as part of the spectral representation of $\widehat{A}_2$. More generally, this means that the value of any physical quantity A whose representing operator $\widehat{A}$ has degenerate eigenvalues may depend on the *context* in which it is considered; in particular, it may depend on the choice of other physical quantities whose associated operators commute with $\widehat{A}$, and which therefore help to label vectors in the degeneracy subspaces of $\widehat{A}$.

Yet another way of expressing the above is to say that the operator $\widehat{A}$ does not represent a *unique* physical quantity: it has a different meaning depending on what other compatible quantities are considered at the same time. This links up with the second possible way of saving value functions: *i.e.*, the idea that the quantisation map $A \mapsto \widehat{A}$ may be many-to-one.

The Kochen–Specker theorem is not directly relevant to the pragmatic approach to quantum theory in so far as no attempt is made in that approach to think of an individual system as possessing values for its observables. However, if this approach is augmented with the assumptions that (i) physical quantities do have values in any individual system; and (ii) these values can be revealed by suitable measurements, then the theorem implies that the results obtained will depend on the context in which an observable is studied. In particular, if $\widehat{A}$, $\widehat{B}$ and $\widehat{C}$ are three operators with $[\widehat{A},\widehat{B}] = 0 = [\widehat{A},\widehat{C}]$, but with $[\widehat{B},\widehat{C}] \neq 0$, then the result of an *individual* measurement of A will depend on whether we choose to measure B or C (or neither) at the same time.

A very important example of this situation arises naturally in the 'EPR' context (see later) of a pair of entangled systems that have become spatially separated, and on which measurements are then made in such a way that the measurement events are space-like separated[4] (in the sense of special relativity). If A is associated with the first system, and B and C with the second, then we expect to be able to measure A simul-

[4]Two events are *space-like* separated if no signal can be sent from one to another without exceeding the speed of light. For any such pair of events, there is always some inertial reference frame with respect to which the two events are simultaneous.

taneously with either B or C, and[5] hence $[\widehat{A}, \widehat{B}] = 0 = [\widehat{A}, \widehat{C}]$. On the other hand, it is easy to arrange that the operators $\widehat{B}$ and $\widehat{C}$ associated with the second system are such that $[\widehat{B}, \widehat{C}] \neq 0$.

Note that the *probabilistic* predictions of the formalism concerning the results of measuring A do not depend on what else is measured at the same time: the probability of getting the result a_m when the state is $|\psi\rangle$ is always just $\langle\psi| \widehat{P}_m |\psi\rangle$. In addition, statements about the hypothetical values of quantities for individual systems are not amenable to experimental tests. In consequence, the deductions above concerning contextuality will not trouble a pragmatist. Indeed, he or she may even support them by citing Bohr's ideas about a property of a quantum system having a meaning only within the context of a specific measurement situation.

However, within a realist interpretation, a contextual assignment of values is in sharp variance with what is normally meant by saying that a property is 'possessed'. In normal discourse, implicit in the statement "the quantity A has a value a" is an understanding that this is so independent of what else might be asserted at the same time. But, according to the Kochen–Specker theorem, in the quantum case we are obliged to talk instead of *pairs* (A, B) or (A, C) having certain values (a, b) and (a', c) *as pairs*, and the implication of contextuality is that a may not equal a'.

This seems particularly odd if the operators $\widehat{B}$ and $\widehat{C}$ that commute with $\widehat{A}$ are associated with measurements made a great distance away. A measurement by a distant observer of B, followed immediately by a measurement of C, would appear to 'cause' an instantaneous jump in A from its value when considered as part of the pair (A, B), to its value when considered as part of the pair (A, C). This is just the type of non-local effect that arises in the context of the Bell inequalities.

Note that the analogue of a value-function in a hidden variable theory would involve the assumption that the value of a physical quantity A in a quantum state $\vec{\psi}$ depends on these additional variables. However, this does not change the force of the arguments above; in particular, it is only in a contextual sense that properties in a hidden variable theory can be said to be 'possessed'.

---

[5]But recall our earlier discussion in Section 6.3 of the idea of two variables being 'simultaneously measurable'.

9.2 The Logic of Quantum Propositions

9.2.1 The Meaning of 'True'

We have alluded several times to the idea that a projection operator $\widehat{P}$ can be regarded as the quantum representative of a proposition P, with the eigenvalues 1 or 0 being associated in some way with P being true or false respectively (equivalently, the proposition is represented by the subspace $\mathcal{H}_{\widehat{P}} := \{\vec{\psi} \in \mathcal{H} \mid \widehat{P}\vec{\psi} = \vec{\psi}\}$ of $\mathcal{H}$). Furthermore, the discussion in the previous section showed that, if it existed, a value function $V_{\vec{\psi}}$ satisfying Eq. (9.2) would associate a number 0 or 1 with each such projector. The obvious interpretation is that $V_{\vec{\psi}}(P) = 1$ and $V_{\vec{\psi}}(P) = 0$ correspond, respectively, to the proposition being true or false when the quantum state is $\vec{\psi}$.

Binary-valued functions of this type (known as *valuations*) play an important role in classical propositional logic, which suggests that it may be profitable to explore quantum theory further from this perspective, particularly the 'logic' of propositions represented by projectors. A central ingredient will presumably be the fact that each (normalised) state $|\psi\rangle$ gives rise to a probability assignment

$$\mathrm{Prob}(P;\vec{\psi}) := \langle\vec{\psi},\widehat{P}\vec{\psi}\rangle = \langle\vec{\psi},\widehat{P}^2\vec{\psi}\rangle = \langle\widehat{P}\vec{\psi},\widehat{P}\vec{\psi}\rangle = \|\,\widehat{P}\vec{\psi}\,\|^2 \tag{9.10}$$

which will be interpreted in some way as the probability that the state of affairs represented by $\widehat{P}$ is 'realised' if the state is $\vec{\psi}$.

Since all propositions in physics can be reduced ultimately to statements about values (possessed, measured, or otherwise) of physical quantities, nothing is lost by focussing on this case. In what follows, we shall discuss only propositions of the type $A \in \Delta$, which asserts that the value of a physical quantity[6] A lies in some subset $\Delta \subset \mathbb{R}$ (or will be found to do so if an appropriate measurement is made; see below). Assertions of this type refer to the state of affairs at some specific time. However, if desired, the formalism can be generalised to include statements about a sequence of physical quantities $A_1, A_2, A_3, \ldots$ whose values are measured (or possessed) at times $t_1, t_2, t_3, \ldots$ with $t_1 < t_2 < t_3 < \cdots$. This leads rather naturally to a quantum version of *temporal* logic, but we shall not discuss this extension here.

[6]It is straightforward to extend the discussion to a finite set of compatible quantities.

Propositions of the type $A \in \Delta$ are what are normally assumed to be represented by spectral projectors.[7] If the spectrum of $\widehat{A}$ is discrete, the operator $\widehat{P}_{A\in\Delta}$ is just the sum of the projectors $\widehat{P}_{A=a_i}$ where the eigenvalues a_i belong to the subset Δ of real numbers. Of course, a special case is when $\Delta = \{a_j\}$ for some specific eigenvalue a_j, in which case $\widehat{P}_{A\in\Delta} = \widehat{P}_{A=a_j}$. The probabilistic predictions of quantum theory are that if the state is $\vec{\psi}$ then $\text{Prob}(A \in \Delta; \vec{\psi}) = \| \widehat{P}_{A\in\Delta}\vec{\psi} \|^2$. In particular, $\text{Prob}(A \in \Delta; \vec{\psi}) = 1$ if and only if $\widehat{P}_{A\in\Delta}\vec{\psi} = \vec{\psi}$, *i.e.*, $\vec{\psi}$ lies in the eigenspace $\mathcal{H}_{A\in\Delta}$ generated by all eigenvectors of $\widehat{A}$ whose eigenvalues are in Δ.[8]

Note that a given projection operator may represent more than one proposition of this type; for example, two operators $\widehat{A}$ and $\widehat{B}$ may share a common spectral projector, so that $\widehat{P}_{A\in\Delta} = \widehat{P}_{B\in\Delta'}$ for some subsets Δ and Δ' of $\mathbb{R}$. In this case, we shall say that the propositions $A \in \Delta$ and $B \in \Delta'$ are *physically equivalent* (but recall the discussion in the preceding section about the possible implications of such a definition if $[\widehat{A}, \widehat{B}] \neq 0$).

Important issues concerning this proposition–projector association include:

- the exact nature of the propositions represented by projection operators;
- the way in which the concepts of 'truth' and 'falsity'[9] apply to such propositions;
- the quantum analogue of the way in which the set of propositions in classical physics acquires the structure of a Boolean algebra from the underlying mathematical representation of states and physical quantities (as discussed in Section 4.2).

---

[7]Every projection operator $\widehat{P}$ lies in the spectral representation of at least one self-adjoint operator; for example, $\widehat{P}$ is self-adjoint and is its own spectral representation!

[8]In what follows, the clumsy notation $\mathcal{H}_{\widehat{P}_T}$ (where T denotes a proposition) will be replaced with the simpler expression $\mathcal{H}_T$.

[9]In any talk of 'truth' or 'falsity' it is important to distinguish between (i) what it means to say that a proposition is true, or false; and (ii) the conditions under which it is justified to make such an assertion. The positivists tended to conflate them, but modern philosophers are more careful. A good, recent discussion of such matters is Grayling (1990).

The nature of the proposition represented by a projection operator depends on the interpretation of quantum theory that is adopted. A realist might want to say that the projection operator $\widehat{P}_{A\in\Delta}$ represents the physical quantity whose value is 1 if the value of A lies in Δ, and is 0 otherwise. Hence, in this case, the proposition $A \in \Delta$ is to be read as

The realist version of $A \in \Delta$:

> "A *has* a value, and this value lies in the subset Δ".

Of course, this leaves open the question of when such an assertion is justified. The situation in classical physics is clear: the proposition is true in a state $s \in \mathcal{S}$ if, and only if, $f_A : \mathcal{S} \to \mathbb{R}$ is such that $f_A(s) \in \Delta$. However, as we saw in the discussion of the Kochen–Specker theorem, any attempt to ascribe possessed values for all quantities in quantum theory can only be maintained at the expense of considering these properties to be contextual; in which case it is not meaningful to say that any particular proposition is either true or false without specifying what other compatible propositions are to be considered at the same time.

One possibility is to adopt the minimal attitude that a physical quantity A can only be said to *have* a value 'a' if (i) a is an eigenvalue[10] of $\widehat{A}$; and (ii) the state $\vec{\psi}$ lies in the associated eigenspace $\mathcal{H}_{A=a}$. One might then argue that, since the first part of the realist version of $A \in \Delta$ is "A has a value, ...", the truth of $A \in \Delta$ requires $|\psi\rangle$ to belong to one of the eigenspaces of $\widehat{A}$; thus $|\psi\rangle$ must be an *eigenvector* of $\widehat{A}$. The second part, "..., and this value lies in Δ", then means that the corresponding eigenvalue must belong to the set $\Delta = \{a_1, a_2, \ldots, a_d\}$. Note that the first requirement excludes any $\vec{\psi}$ that is a *sum* of elements in more than one of the eigenspaces $\mathcal{H}_{A=a_1}, \mathcal{H}_{A=a_2}, \ldots, \mathcal{H}_{A=a_d}$, even though the formalism gives $\mathrm{Prob}(A \in \Delta; \vec{\psi}) = 1$ for all such states. In these circumstances, it is clearly inappropriate to regard the spectral projector $\widehat{P}_{A\in\Delta}$ as representing the proposition $A \in \Delta$. Therefore, if one wanted to maintain the proposition–projector link, it would be natural to adopt the weaker condition that $A \in \Delta$ is true in a state $\vec{\psi}$ if $\widehat{P}_{A\in\Delta}\vec{\psi} = \vec{\psi}$, *i.e.*, if $\vec{\psi}$ is any linear combination of eigenvectors of $\widehat{A}$ whose corresponding eigenvalues lie in Δ.

---

[10] This requires careful rephrasing if the spectrum of $\widehat{A}$ includes continuous pieces.

The situation is less ambiguous in the instrumentalist approaches to quantum theory. The projector $\widehat{P}_{A\in\Delta}$ now represents a physical observable that is defined operationally by specifying how it is to be measured: *i.e.*, measure A, and then, if the result lies in Δ, assign the number 1 to the observable, otherwise set it equal to 0. The proposition $A \in \Delta$ should now be read as asserting that a *measurement* of A will yield a result that lies in the subset Δ.

On the face of it, the statement "if a measurement of A is made, then the result will lie in Δ" is applicable to a single system, as is the realist claim that "A has a value that lies in Δ". However, the instrumentalist form of the proposition is not a statement about how things are, but rather a claim about what *would* happen *if* a certain operation was performed. This seems to cry out for a positivist-type verification: *i.e.*, make the measurement and see what happens—if the result does not lie in Δ, then the proposition is certainly false.

But what if the result *does* lies in Δ: does this mean the proposition is true? Certainly not if, as is arguably the case, a more precise rendering of the proposition is

The modal version of $A \in \Delta$:

> "if a measurement of A is made, then, *necessarily*, the result will lie in the subset Δ"

Put in this modal[11] form, the proposition is manifestly counterfactual[12] and cannot be verified by any single, or finite set, of measurements. Indeed, it could be argued that, in this form, the proposition does not apply at all to an individual system but only to a collection of such on which repeated measurements are to be made. Thus, in this reading, it is consistent to say that states and propositions apply only to such collections.

The discussion above shows why, in the relative-frequency, anti-realist interpretations of quantum theory, we are lead naturally to read the proposition $A \in \Delta$ as

The probability-one version of $A \in \Delta$:

---

[11]For an extensive discussion of modal concepts in quantum theory see Van Frassen (1991).

[12]A general discussion of the general significance of counterfactual statements in quantum theory is contained in Peres (1993).

> "if a measurement of A is made, then, *with probability one*, the result will lie in Δ",

in which case it is clear when it would be asserted as true. Namely:

> "if the quantum state is $\vec{\psi}$, a proposition T represented by a projection operator $\widehat{P}_T$ is true if $\text{Prob}(T;\vec{\psi}) = 1$".

Comments

1. Since $\text{Prob}(T;\vec{\psi}) = \parallel \widehat{P}_T\vec{\psi} \parallel^2$, the condition $\text{Prob}(T;\vec{\psi}) = 1$ implies that [Exercise] $\widehat{P}_T\vec{\psi} = \vec{\psi}$, *i.e.*, $\vec{\psi}$ is in fact an eigenvector of $\widehat{P}_T$ with eigenvalue 1.

2. This result implies that the proposition $A \in \Delta$ is true in any state $\vec{\psi}$ that can be written as a linear *superposition* of the eigenvectors associated with eigenvalues in Δ. Of course, this does *not* mean that the probability of getting any *specific* eigenvalue is equal to 1; indeed, this number will be strictly less than 1 if $\vec{\psi}$ has non-zero expansion coefficients for more than one of the eigenspaces. Note that, unlike in the realist case, this causes no problems, since the modal form of $A \in \Delta$ does not say that a measurement of A necessarily gives any one *particular* element of Δ.

3. Care must be taken when handling propositions in this modal form. For example, if the proposition $A = a_m$ is interpreted as "if A is measured then the result will necessarily be a_m", then the value of $\text{Prob}(A = a_m;\vec{\psi})$ cannot be read as the probability that the proposition is true when the state is $\vec{\psi}$. Rather, $\text{Prob}(A = a_m;\vec{\psi})$ refers to the relative frequency with which the result a_m will be found on the given collection of systems. For example, the statement "there is a 70% chance that the proposition 'if A is measured, then the result will necessarily be a_m' is true", is quite different from "if A is measured, then there is a 70% chance that the result will be a_m". It is the latter that is intended in the quantum formalism by the statement $\text{Prob}(A = a_m;\vec{\psi}) = 0.7$. Note that, if desired, one might introduce an extended class of modal propositions of the form

> "if a measurement of A is made, then the probability that the result will lie in Δ is r",

where the real number r satisfies $0 \leq r \leq 1$. The special value $r = 1$ then corresponds to the probability-one proposition discussed above. However, we shall not pursue this option here.

9.2.2 Is 'False' the Same as 'Not True'?

The next issue to consider is what it means to say that a proposition is *false*, and under what circumstances in quantum theory such an assertion would be justified. In classical physics or logic, this causes no difficulties: a proposition is false if, and only if, it is not true.

If we adopt this idea in quantum theory then, in the relative-frequency interpretations, "$A \in \Delta$ is false" means that, in a long series of repeated measurements of A, one or more numbers not lying in Δ will occur with a non-zero relative frequency. In general, this suggests the definition that a proposition T is 'false' in a state $\vec{\psi}$ if $\text{Prob}(T; \vec{\psi}) < 1$.

On the other hand, a realist may balk at the idea that the proposition $A \in \Delta$ could be said to be false if there is still a non-zero probability of 'finding' values in Δ. When confronted with the existing formalism of quantum theory—in particular the idea that propositions are to be associated with projection operators—a realist may feel that the most natural ascription of truth and falsity in a state $\vec{\psi}$ is that a proposition $A \in \Delta$ is (i) true in a state $\vec{\psi}$ if $\text{Prob}(A \in \Delta; \vec{\psi}) = 1$ (*i.e.*, $\vec{\psi} \in \mathcal{H}_{A\in\Delta}$); and (ii) false if $\text{Prob}(A \in \Delta; \vec{\psi}) = 0$ (*i.e.*, $\vec{\psi} \in \mathcal{H}^{\perp}_{A\in\Delta}$).

However, any vector $\vec{\psi} \in \mathcal{H}$ can be written as a sum of components in $\mathcal{H}_{A\in\Delta}$ and $\mathcal{H}^{\perp}_{A\in\Delta}$, and, according to these ascriptions, the proposition $A \in \Delta$ cannot be said to be either true *or* false if both components are non-zero. This suggests that some sort of multi-valued logic should be used and, indeed, several attempts of this type have been made. The simplest approach, proposed first by Reichenbach (1944) (see also Putnam (1957)), is to use a three-valued logic in which a proposition can be either true, false, or indeterminate.[13] The typical quantum-mechanical situation in which a non-trivial superposition of eigenvectors leads to a proposition

[13]Rather as in Scotland, a jury can return a verdict of guilty, not guilty, or not proven. Of course, there it is assumed that the accused is, in fact, either guilty or not guilty: the implications of the Kochen–Specker theorem have not yet found their way into the legal system!

being neither true nor false would then be assigned to the indeterminate category.

A more complex possibility is to use a continuous spectrum of truth values lying between 0 and 1, with the truth value of the proposition T in a state $\vec{\psi}$ being set equal to $\text{Prob}(T; \vec{\psi})$; this is the idea that lies behind the subject of *fuzzy* logic. But, in either case, it is not really clear what the meaning is of a proposition that is deemed to apply to a single system but which is then said to be neither true nor false. Philosophers have debated this for some time, but no definitive conclusion has emerged. However, serious worries have been expressed about the idea of using a multi-valued logic within a mathematical formalism that is, itself, based firmly on classical two-valued logic.

Another major problem with multi-valued logics is to decide on the appropriate analogues of the usual logical connectives 'and', 'or', 'negation' and 'logical implication' that were introduced in Section 4.2 in the context of classical physics. A variety of attempts has been made at a purely mathematical level, but no clear consensus has emerged.

The difficulties with logical connectives can be alleviated by adopting the strategy in which a proposition $A \in \Delta$ is understood to be false in a state $\vec{\psi}$ if it is not true, *i.e.*, if $\text{Prob}(A \in \Delta; \vec{\psi}) < 1$. This approach is quite natural in instrumentalist interpretations of quantum theory, and results in a two-valued propositional structure, albeit at the expense of introducing a certain asymmetry between the concepts of 'true' and 'false'. Of course, it is still necessary to define the logical connectives.

9.2.3 The Logical Connectives

The construction of a 'quantum logic' on the set of two-valued (true/false) propositions was first discussed long ago in a famous paper by Birkhoff & von Neumann (1936) that stimulated a research programme that has been the subject of considerable interest ever since. They started with the set $\mathcal{P}(\mathcal{H})$ of all closed subspaces of a Hilbert space $\mathcal{H}$ (or, equivalently, with the set of all projection operators) and suggested certain analogues of the classical logical operations. The basic definitions are as follows.

1. A proposition T *implies* another U, written $T \preceq U$, if for all states $\vec{\psi}$ such that $\text{Prob}(T; \vec{\psi}) = 1$ we have $\text{Prob}(U; \vec{\psi}) = 1$; *i.e.*, U is true in all

states in which T is true. This is equivalent to $\mathcal{H}_T$ being a closed *subspace* of $\mathcal{H}_U$. The analogous classical relation is $\mathcal{S}_T \subset \mathcal{S}_U$ (see Section 4.3).

If $\widehat{P}_T$ and $\widehat{P}_U$ are the associated projection operators, the relation $T \preceq U$ is equivalent to

$$\widehat{P}_T \widehat{P}_U = \widehat{P}_U \widehat{P}_T = \widehat{P}_T, \tag{9.11}$$

which is denoted $\widehat{P}_T \preceq \widehat{P}_U$.

Note that if T implies U and U implies T, it is natural to say that the propositions T and U are equivalent. Indeed, the former means that $\widehat{P}_T \widehat{P}_U = \widehat{P}_T$ and $\widehat{P}_T \widehat{P}_U = \widehat{P}_U$, *i.e.*, $\widehat{P}_T = \widehat{P}_U$, which is the definition of physical equivalence suggested earlier. As is the case in classical physics, the appropriate objects to study are the *equivalence classes* of propositions which, in the quantum case, are therefore the projection operators. Thus it is more appropriate to view the logical operations as defined on projection operators (or, equivalently, topologically closed subspaces), than on propositions *per se*.

2. The operator that corresponds to the *identically false* proposition is the null operator $\widehat{0}$ that maps every vector in $\mathcal{H}$ to the null vector $\vec{0}$, and the operator that corresponds to the *identically true* proposition is the unit operator $\widehat{\mathbb{1}}$. Thus, for any projection operator $\widehat{P}$, we have $\widehat{0} \preceq \widehat{P} \preceq \widehat{\mathbb{1}}$. In classical physics, the analogues of these two propositions are the empty set $\emptyset$ and the whole space $\mathcal{S}$ respectively. And, of course, for all subsets $\mathcal{S}_T$ of $\mathcal{S}$ we have $\emptyset \subset \mathcal{S}_T \subset \mathcal{S}$.

3. If T is a proposition, the proposition $\neg T$ (*not* T) is defined by requiring that, for all states $\vec{\psi}$, $\text{Prob}(\neg T; \vec{\psi}) = 1$ if, and only if, $\text{Prob}(T; \vec{\psi}) = 0$. This is equivalent to the relation $\mathcal{H}_{\neg T} = (\mathcal{H}_T)^{\perp}$. This relation determines a unique projection operator $\widehat{P}_{\neg T}$ that represents $\neg T$:

$$\widehat{P}_{\neg T} = \widehat{\mathbb{1}} - \widehat{P}_T \tag{9.12}$$

which actually implies the (superficially stronger) relation between the probabilities of T and $\neg T$

$$\text{Prob}(T; \vec{\psi}) + \text{Prob}(\neg T; \vec{\psi}) = 1 \tag{9.13}$$

for all states $\vec{\psi}$. The classical analogue of Eq. (9.12) is $\mathcal{S}_{\neg T} = \mathcal{S} - \mathcal{S}_T$, or Eq. (4.17).

Unlike the classical case, the quantum definition has the peculiar property that T can be *false* in a state $\vec{\psi}$ (*i.e.*, T is not true, so that $\text{Prob}(T;\vec{\psi}) < 1$) *without* implying that $\neg T$ is *true* (which requires $\text{Prob}(T;\vec{\psi}) = 0$). But, as expected, $\neg T$ is true *does* imply that T is false! This curious asymmetry stems directly from (i) the existence of linear superpositions of eigenstates of the operator $\widehat{P}_T$; and (ii) our desire to have a binary-valued, rather than trinary-valued, logic.

4. Let T and U be a pair of propositions with associated closed subspaces $\mathcal{H}_T$ and $\mathcal{H}_U$, and projection operators $\widehat{P}_T$ and $\widehat{P}_U$, respectively. Then the proposition $T \wedge U$ (to be interpreted as some type of conjunction, T *and* U) is defined by requiring that $\text{Prob}(T \wedge U;\vec{\psi}) = 1$ in a state $\vec{\psi}$ if, and only if, $\text{Prob}(T;\vec{\psi}) = 1$ and $\text{Prob}(U;\vec{\psi}) = 1$ *i.e.*, $T \wedge U$ is true in a state $\vec{\psi}$ if, and only if, both T and U are true. This means

$$\mathcal{H}_{T\wedge U} = \mathcal{H}_T \bigcap \mathcal{H}_U \tag{9.14}$$

which is reminiscent of the classical result $\mathcal{S}_{T\wedge U} = \mathcal{S}_T \bigcap \mathcal{S}_U$ (*cf.* Eq. (4.15)).

Expressed in terms of subspaces, the quantum definition seems rather natural. However, there is no simple way of translating it into the language of projection operators since the operator $\widehat{P}_T \wedge \widehat{P}_U := \widehat{P}_{T\wedge U}$ that projects onto $\mathcal{H}_T \bigcap \mathcal{H}_U$ is not a simple function of the projection operators $\widehat{P}_T$ and $\widehat{P}_U$. If $[\,\widehat{P}_T, \widehat{P}_U\,] = 0$ it can be shown that the projection operator $\widehat{P}_T \wedge \widehat{P}_U$ that represents the proposition $T \wedge U$ is

$$\widehat{P}_T \wedge \widehat{P}_U = \widehat{P}_T \widehat{P}_U. \tag{9.15}$$

Otherwise, all that can be shown is

$$\widehat{P}_T \wedge \widehat{P}_U = \lim_{k\to\infty} (\widehat{P}_T \widehat{P}_U)^k \tag{9.16}$$

where the limit uses an appropriate notion of convergence of a sequence of bounded operators.

5. In classical physics, the set that represents the disjunction $T \vee U$ (T *or* U) of a pair of propositions T and U, is the union $\mathcal{S}_T \bigcup \mathcal{S}_U$ of the sets $\mathcal{S}_T$ and $\mathcal{S}_U$ that represent them, *i.e.*, it is the smallest subset of the state space $\mathcal{S}$ that contains both $\mathcal{S}_T$ and $\mathcal{S}_U$.

The obvious analogue in quantum theory is to set $\mathcal{H}_{T\vee U}$ equal to the union of $\mathcal{H}_T$ and $\mathcal{H}_U$. But this cannot be correct since $\mathcal{H}_T \bigcup \mathcal{H}_U$ is not a *linear* subspace of $\mathcal{H}$, *i.e.*, it is not closed under the operation of taking sums of vectors. This suggests that a more appropriate definition would be $\mathcal{H}_{T\vee U} := \mathcal{H}_T + \mathcal{H}_U$, which is defined to be the linear sum of all vectors in $\mathcal{H}_T$ and $\mathcal{H}_U$. However, even this is not correct if $\mathcal{H}$ has an infinite dimension since then $\mathcal{H}_T + \mathcal{H}_U$ may not be topologically closed. This defect is remedied by the final definition

$$\mathcal{H}_{T\vee U} := \overline{\mathcal{H}_T + \mathcal{H}_U} \tag{9.17}$$

which is the smallest, topologically closed linear subspace of $\mathcal{H}$ that contains both $\mathcal{H}_T$ and $\mathcal{H}_U$.

As in the case of the conjunction, there is no easy way to write the definition Eq. (9.17) in terms of the associated projection operators. However, if $[\widehat{P}_T, \widehat{P}_U] = 0$ then the projection operator $\widehat{P}_T \vee \widehat{P}_U := \widehat{P}_{T\vee U}$ that represents the proposition $T \vee U$ is

$$\widehat{P}_T \vee \widehat{P}_U = (\widehat{P}_T + \widehat{P}_U - \widehat{P}_T\widehat{P}_U). \tag{9.18}$$

In particular, if $\widehat{P}_T\widehat{P}_U = 0$ then $\widehat{P}_T \vee \widehat{P}_U = \widehat{P}_T + \widehat{P}_U$ (*cf.* the classical result Eq. (4.16)).

Comments

1. The closed subspaces $\mathcal{H}_T$ and $\mathcal{H}_U$ of $\mathcal{H}$ satisfy the equation

$$\overline{\mathcal{H}_T + \mathcal{H}_U} = (\mathcal{H}_T^{\perp} \bigcap \mathcal{H}_U^{\perp})^{\perp}. \tag{9.19}$$

This is an analogue of the DeMorgan rule of classical set theory

$$\mathcal{S}_T \bigcup \mathcal{S}_U = \mathcal{S} - \Big((\mathcal{S} - \mathcal{S}_T)\bigcap(\mathcal{S} - \mathcal{S}_U)\Big) \tag{9.20}$$

which corresponds to the logical identity $T \vee U = \neg(\neg T \wedge \neg U)$. This provides a further justification for the consistency of the above assignment of logical connectives.

2. Care must be taken in interpreting the assignment of logical connectives. This is particularly true of the idea of '$T \wedge U$': one may interpret $\wedge$ as 'and' in the normal sense only if the two operators commute since, as mentioned earlier, only then[14] is it unequivocally meaningful to assert

[14]This is not quite correct. Simultaneous eigenvectors can exist for a pair of operators that do not commute. However, these vectors will necessarily be too few in number to form a basis set.

that states of the system exist in which both properties are possessed at once. In the non-commuting case, Eq. (9.16) is sometimes read as saying that $T \wedge U$ refers to the results of an infinite sequence of measurements of T and U.

3. The fact that most pairs of operators do *not* commute is reflected in the crucial property of the set of quantum propositions that the algebraic structure associated with the definitions above of $\wedge$ and $\vee$ fails to be distributive.

This can be seen easily with the aid of a simple example. Let $\mathcal{H}_{\psi_1}$ and $\mathcal{H}_{\psi_2}$ be one-dimensional subspaces of a two-dimensional Hilbert space $\mathcal{H}$, spanned by (*i.e.*, all complex multiples of) the linearly-independent vectors $\vec{\psi}_1$ and $\vec{\psi}_2$ respectively, and let $\vec{\phi}$ be any non-trivial linear combination of $\vec{\psi}_1$ and $\vec{\psi}_2$. Then $\mathcal{H}_\phi \bigcap (\mathcal{H}_{\psi_1} + \mathcal{H}_{\psi_2})$ is simply $\mathcal{H}_\phi$ (because $\mathcal{H}_{\psi_1} + \mathcal{H}_{\psi_2} = \mathcal{H}$), whereas $(\mathcal{H}_\phi \bigcap \mathcal{H}_{\psi_1}) + (\mathcal{H}_\phi \bigcap \mathcal{H}_{\psi_2})$ is the null vector, since the subspaces on each side of the $+$ sign are null. This contrasts sharply with the distributive property of classical propositions as evidenced in Eq. (4.18) and which, we recall, is a direct consequence of the Boolean nature of the algebra of intersections and unions of subsets of a classical state space.

Most of the strange, non-classical features of quantum theory can be traced back to this non-distributive property of the logical structure of quantum propositions—an observation that motivates the full *quantum logic* approach to quantum theory. This starts by noting the non-distributive structure of the set $\mathcal{P}(\mathcal{H})$ of all projection operators on a Hilbert space $\mathcal{H}$, which is regarded as a quantum analogue of the Boolean logic of classical physics. The idea then is to abstract out this property to form a new approach to quantum theory in which the basic objects are not a Hilbert space of states and its associated self-adjoint operators, but rather a non-distributive algebra of putative propositions, and probability measures defined on this algebra.

A vast literature has accrued on this topic that throws much light on the underlying structure of quantum theory, and on the possibility of going beyond the standard Hilbert space formalism to create a new type of theory. It also plays an important role in discussions of realist approaches to quantum theory in which the propositions concerned are interpreted in terms of possessed properties rather than results of repeated measurements. Good introductions to this area include Jauch (1973), and Hughes (1989); a more comprehensive account is Beltrametti & Cassinelli (1981).

9.2.4 Gleason's Theorem

The probabilistic rules for density matrices discussed in Section 6.1 imply that for any proposition P, with associated projector $\widehat{P} \in \mathcal{P}(\mathcal{H})$, the probability associated with P in the state $\widehat{\rho}$ is

$$\text{Prob}(P;\rho) = \text{tr}\,(\widehat{\rho}\,\widehat{P}). \tag{9.21}$$

This raises an interesting and important question: are there any other ways of assigning probabilities to the elements of $\mathcal{P}(\mathcal{H})$, or is the standard quantum formalism unique? More precisely, can we find a new probabilistic theory by starting with some Hilbert space $\mathcal{H}$, with its non-distributive algebra $\mathcal{P}(\mathcal{H})$ of subspaces/projection operators, and then construct a probability map $\text{Prob} : \mathcal{P}(\mathcal{H}) \to \mathbb{R}$ that is *not* of the type in Eq. (9.21)? In particular, what is the space of states in such a theory? This was one of the most important issues addressed in the early studies of quantum logic. The hope was that non-standard theories (particularly those involving hidden variables) could be constructed in this way.

To tackle this question we must decide first what requirements should be satisfied by a 'probability map' $\text{Prob} : \mathcal{P}(\mathcal{H}) \to \mathbb{R}$. The analogous question in classical statistical physics was posed briefly in Section 4.3 where a general 'probability measure' was defined to be a real-valued function μ of subsets of $\mathcal{S}$ that satisfies the three conditions Eq. (4.12–4.14).

In the quantum case, one obvious condition is $0 \leq \text{Prob}(P) \leq 1$ for all propositions $P \in \mathcal{P}(\mathcal{H})$; two other, essentially trivial, requirements are $\text{Prob}(\emptyset) = 0$ and $\text{Prob}(\mathbb{1}) = 1$, where $\emptyset$ and $\mathbb{1}$ correspond respectively to the propositions that are identically false and true. The key question is what should be required of the probabilities $\text{Prob}(T \wedge U)$ and $\text{Prob}(T \vee U)$ for a pair of propositions T and U.

It is helpful to start by considering the analogous question in classical physics, where, as discussed in Section 4.2, propositions correspond to subspaces of the state space $\mathcal{S}$. In this case, bearing in mind that $\mathcal{S}_{T\wedge U} =$

$\mathcal{S}_T \bigcap \mathcal{S}_U$ and $\mathcal{S}_{T\vee U} = \mathcal{S}_T \bigcup \mathcal{S}_U$, it is easy to see with the aid of a Venn diagram that one required condition is

$$\text{Prob}(T \vee U) = \text{Prob}(T) + \text{Prob}(U) - \text{Prob}(T \wedge U). \tag{9.22}$$

We should not expect to impose this condition in the quantum case because of the problem of incompatible observables. However, a special case of Eq. (9.22) is when the subspaces $\mathcal{S}_T$ and $\mathcal{S}_U$ are disjoint (*i.e.*, $\mathcal{S}_T \bigcap \mathcal{S}_U = \emptyset$), so that the corresponding propositions are mutually exclusive. In this case, Eq. (9.22) becomes the *finite additivity* condition

$$\text{Prob}(T \vee U) = \text{Prob}(T) + \text{Prob}(U) \tag{9.23}$$

which, if $\{P_1, P_2, \ldots, P_M\}$ is any finite set of propositions that are pairwise exclusive, generalises at once to

$$\text{Prob}(P_1 \vee P_2 \cdots \vee P_M) = \sum_{i=1}^{M} \text{Prob}(P_i). \tag{9.24}$$

An important technical requirement in classical probability theory is *countable additivity*, *i.e.*, the condition Eq. (9.24) should continue to hold in the limit $M \to \infty$.

The quantum analogue of two mutually-exclusive propositions is a pair of projection operators $\widehat{P}_T, \widehat{P}_U$ that are orthogonal, *i.e.*, $\widehat{P}_T\widehat{P}_U = 0 = \widehat{P}_U\widehat{P}_T$. In this case, problems of incompatibility do not apply, and it is natural to suppose that Eq. (9.24) passes across in the form $\widehat{P}_1 \vee \widehat{P}_2 \cdots \vee \widehat{P}_M = \sum_{i=1}^{M} \widehat{P}_i$ for any collection of pairwise orthogonal projection operators. As in the classical case, it may be desirable to extend this requirement to the limit $M \to \infty$. Thus, our final conclusion is that a minimal set of requirements on any quantum probability function $\text{Prob} : \mathcal{P}(\mathcal{H}) \to \mathbb{R}$ is

$$0 \leq \text{Prob}(\widehat{P}) \leq 1 \text{ for all } \widehat{P} \in \mathcal{P}(\mathcal{H}) \tag{9.25}$$

$$\text{Prob}(\widehat{0}) = 0, \qquad \text{Prob}(\widehat{1}) = 1 \tag{9.26}$$

$$\text{Prob}(\sum_{i=1}^{\infty} \widehat{P}_i) = \sum_{i=1}^{\infty} \text{Prob}(\widehat{P}_i) \tag{9.27}$$

for any finite or countably infinite set $\{\widehat{P}_1, \widehat{P}_2, \ldots\}$ of projection operators that are pairwise mutually orthogonal. Typically, such sets arise as the spectral projectors of a self-adjoint operator.

As we saw in Section 6.1, one way of satisfying Eq. (9.25–9.27) is to define

$$\mathrm{Prob}(\widehat{P}) := \mathrm{tr}\,(\widehat{\rho}\,\widehat{P}) \tag{9.28}$$

where $\widehat{\rho}$ is any density matrix. A remarkable theorem due to Gleason (1957) shows that, provided $\dim \mathcal{H} > 2$, the converse is true: the *only* way of satisfying Eq. (9.25–9.27) is with the aid of a density matrix, and Eq. (9.28). The proof is geometrical in nature, and there are a number of technical links between Gleason's result and the theorem of Kochen and Specker.

Gleason's theorem places strong constraints on any attempts to modify the standard quantum formalism as, for example, might be desired in the construction of hidden variable theories. It also gives a rather deep reason why density matrices play such an important role in quantum theory.

9.3 Non-Locality and the Bell Inequalities

9.3.1 EPR and the Incompleteness of Quantum Theory

The second major problem confronting hidden variables and possessed properties was first spelt out in a famous paper by Einstein, Podolsky & Rosen (1935). Their original analysis was rather complex in a technical sense, and most discussion these days uses a simpler version due to Bohm (1951). He considered a particle whose decay produces two spin-$\frac{1}{2}$ particles whose total spin angular momentum is zero. These particles move away from each other in opposite directions, and the components of their spins along various directions are subsequently measured by two observers, N and L say. The constraint on the total spin means that if both observers agree to measure the spin along a particular direction $\mathbf{n}$, and if N measures $+\frac{1}{2}\hbar$, then L will necessarily get the result $-\frac{1}{2}\hbar$, and if N measures $-\frac{1}{2}\hbar$, then L will necessarily get $\frac{1}{2}\hbar$.

There are no surprises if such correlations are analysed in the context of classical physics. If one particle emerges from the decay with its internal angular momentum vector pointing along some particular direction then, because of conservation of angular momentum, the second particle is guaranteed to emerge with its spin vector pointing in the opposite direction. Thus the 100% anti-correlations found in the measurements made by the two observers are simply the result of the fact that both particles *possess* actual, and (anti-)correlated, values of internal angular momentum; and this is true from the time they emerge from the decay to the time the measurements are made. There are no paradoxes here, and everything is in accord with the simple realist view of classical physics.

The situation in quantum theory is radically different. Suppose first that the measurements are made along the z-axes of the two observers. The spin part of the state of the two particles can be written in terms of the associated eigenvectors as Eq. (8.46)

$$|\psi\rangle = \frac{1}{\sqrt{2}}(|\uparrow\rangle|\downarrow\rangle - |\downarrow\rangle|\uparrow\rangle) \tag{9.29}$$

where, for example, $|\uparrow\rangle|\downarrow\rangle$ is the state in which particles 1 and 2 have spin $+\frac{1}{2}\hbar$ and $-\frac{1}{2}\hbar$ respectively. Thus $\widehat{S}_z|\uparrow\rangle = \frac{1}{2}\hbar|\uparrow\rangle$ and $\widehat{S}_z|\downarrow\rangle = -\frac{1}{2}\hbar|\downarrow\rangle$.

The pragmatic or instrumentalist interpretation of the entangled state $|\psi\rangle$ is straightforward: if, in a series of repeated measurements by N, a selection is made of the pairs of particles for which the measurement of her particle gave spin-up, then—with probability one—a series of measurements by L on her particle in these pairs will yield spin-down. Similarly, if N finds spin-down then, with probability one, L will find spin-up. This correlation can be explained by saying that the measurements by N (computed with the operator $\widehat{S}_z \otimes \widehat{\mathbb{1}}$) 'cause' a reduction of the state vector from $|\psi\rangle$ to $|\uparrow\rangle|\downarrow\rangle$ or $|\downarrow\rangle|\uparrow\rangle$ respectively according to whether the spin-up or spin-down result is selected. This new state is an eigenstate of the operator $\widehat{\mathbb{1}} \otimes \widehat{S}_z$ associated with the second particle, and with an eigenvalue that is the opposite of the result obtained by N.

This description is acceptable within the confines of the pragmatic approach to quantum theory, but difficulties arise if one tries to enforce a more realist interpretation of the entangled state in Eq. (9.29). The obvious question is how the information about each observer's individual results 'gets to' the other particle to guarantee that the result obtained by the second observer will be the correct one.

One might be tempted to invoke the reasoning of classical physics and argue that both particles *possess* the appropriate value all the time.

However, the only way in standard quantum theory of guaranteeing that a certain result will be obtained is if the state is an eigenvector of the observable concerned. But the state $|\psi\rangle$ in Eq. (8.46) is *not* of this type: rather it displays the typical features of quantum entanglement and is in fact a superposition of such states. And any attempt to invoke a hidden variable resolution will have to cope with the implications of the Kochen–Specker theorem. There is also a question of whether this picture is compatible with special relativity. If the measurements by the two observers are space-like separated (which could easily be arranged) then which of them makes the first measurement—and hence, in the standard interpretation, causes the state-vector reduction—is reference-frame dependent.

The problem is compounded by considering what happens if the observers decide to measure, say, the x component of the spins, rather than the z component. The state in Eq. (9.29) can now be written in terms of the S_x eigenvectors in Eq. (8.50) as

$$|\psi\rangle = \frac{1}{\sqrt{2}}(|\leftarrow\rangle|\rightarrow\rangle - |\rightarrow\rangle|\leftarrow\rangle) \tag{9.30}$$

where $|\rightarrow\rangle$ and $|\leftarrow\rangle$ correspond to eigenvalues $+\frac{1}{2}\hbar$ and $-\frac{1}{2}\hbar$ respectively of the operator $\widehat{S}_x$.

In one sense, Eq. (9.30) is what might have been expected, and confirms that there is the same type of 100% anti-correlation between the S_x measurements as that found for the observable S_z. Indeed, this argument can be generalised to show that for *any* unit vector $\mathbf{n}$, Eq. (9.29) can be rewritten as the sum of two anti-correlated terms containing eigenvectors of the projection $\mathbf{n}\cdot\widehat{\mathbf{S}}$ of the spin operator along $\mathbf{n}$. Thus, if one adopts the classical type of argument, one is obliged to conclude that both particles possessed exact values of spin along *any* axis from the moment they left the decay. This might not be easy to reconcile with the uncertainty relations associated with the angular momentum commutators.

Einstein, Podolsky and Rosen considered these issues, and concluded that the difficulties could be resolved in one of only two ways:

1. When N makes her measurement, the result communicates itself at once in some way to particle 2, and converts its state into the appropriate eigenvector. Or,

2. quantum theory is incomplete and provides only a partial specification of the actual state of the system.

In contemplating the first possibility it must be appreciated that the two particles may have moved a vast distance apart before the first measurement is made, and therefore any 'at once' mode of communication would be in violent contradiction with the spirit (if not the law) of special relativity. It is not surprising that Einstein was not very keen on this alternative! An additional objection involves the lack within quantum theory itself of any idea about how this non-local effect is supposed to take place, so in this sense the theory would be incomplete anyway.

Einstein, Podolsky and Rosen came to the conclusion that the theory is indeed incomplete, although they left open the question of the correct way in which to 'complete' it.[15] One natural path is to suppose that there exist 'hidden variables' whose values are not accessible to measurement in the normal way but which determine the actual values of what we normally regard as observables—rather as do the microstates in classical statistical physics. However, it is not a trivial matter to construct a hidden variable theory that reproduces the empirical results of quantum mechanics. In particular, such a theory would need to explain why it is that certain observables are incompatible (those with a non-vanishing commutator) in the sense that one cannot prepare a state of the system that violates the predictions of the uncertainty relations. In addition, there is a need to come to terms with the implications of the Kochen–Specker theorem.

Hidden variable theories capable of reproducing the results of conventional quantum theory do in fact exist[16] but they exhibit a non-locality which is every bit as peculiar as that discussed above. One might think this was merely a deficiency of these particular theories, and that others might exist without this problem. However, a very famous result of John Bell shows that this is not possible. That is, any hidden-variable theory that exactly replicates the results of quantum theory will necessarily possess striking non-local features. This result, which is of major importance

[15]For a subtle analysis of Einstein's views of realism and quantum theory see Fine (1986).

[16]David Bohm invented a famous example. For a stimulating and comprehensive summary of his approach to quantum theory see Bohm & Hiley (1993). Another very readable recent account is Holland (1993).

in appreciating the conceptual challenge posed by quantum theory, forms the content of the next, and final, section of these notes.

9.3.2 The Bell Inequalities

As with the Kochen–Specker result, the non-locality property we are about to discuss is not just a feature of hidden variable theories: it applies to *any* realist interpretation of quantum theory in which it is deemed meaningful to say that an individual system *possesses* values for its physical quantities in a way that is analogous to that in classical physics. We shall derive an inequality that is satisfied by certain correlation functions in any such theory which is also local. We shall then see how the predictions of quantum theory can violate this inequality. Since Bell's original work, many examples of 'Bell inequalities' have been discovered, and the one employed here is chosen because of its simplicity (see Redhead (1989) for more discussion of this particular example).

The considerations of EPR were concerned with two observers who make measurements along the same axis. John Bell found his famous inequalities by asking what happens if the observers measure the spin of the particles along different axes. In particular, we consider a pair of unit vectors $\mathbf{a}$ and $\mathbf{a}'$ for one observer, and another pair $\mathbf{b}$ and $\mathbf{b}'$ for the other. Now suppose a series of repeated measurements is made on a collection of systems whose quantum state is described by the vector $|\psi\rangle$ in Eq. (9.29); for example we could look at a series of decays, each of which produces a pair of particles with zero total spin angular momentum. The central realist assumption we are testing is that each particle has a definite value at all times for any direction of spin. We let a_n denote $2/\hbar$ times the value of $\mathbf{a}\cdot\mathbf{S}$ possessed by particle 1 in the n'th element of the collection. Thus $a_n = \pm 1$ if $\mathbf{a}\cdot\mathbf{S} = \pm\frac{1}{2}\hbar$.

The key ingredient in the derivation of the Bell inequalities is the correlation between measurements made by the two observers along these different directions. For directions $\mathbf{a}$ and $\mathbf{b}$ this is defined by

$$C(\mathbf{a},\mathbf{b}) := \lim_{N\to\infty} \frac{1}{N}\sum_{n=1}^{N} a_n b_n, \tag{9.31}$$

and similarly for the other directions. Note that if the results are always totally correlated then $C(\mathbf{a}, \mathbf{b}) = +1$, whereas if they are totally anti-correlated we get $C(\mathbf{a}, \mathbf{b}) = -1$.

Now look at the quantity

$$g_n := a_n b_n + a_n b'_n + a'_n b_n - a'_n b'_n. \tag{9.32}$$

For any member n of the collection, each term in this sum will take on the value $+1$ or -1. Furthermore, the fourth term on the right hand side is equal to the product of the first three (because $(a_n)^2 = 1 = (b_n)^2$). Then thinking about the various possibilities shows that g_n can take on only the values ± 2. Therefore, the right hand side of the expression

$$\left| \frac{1}{N} \sum_{n=1}^{N} g_n \right| = \left| \frac{1}{N} \sum_{n=1}^{N} a_n b_n + \frac{1}{N} \sum_{n=1}^{N} a_n b'_n + \frac{1}{N} \sum_{n=1}^{N} a'_n b_n - \frac{1}{N} \sum_{n=1}^{N} a'_n b'_n \right|, \tag{9.33}$$

representing the average value of g_n, must be less than or equal to 2. Thus, in the limit as $N \to \infty$, we get

$$|C(\mathbf{a}, \mathbf{b}) + C(\mathbf{a}, \mathbf{b}') + C(\mathbf{a}', \mathbf{b}) - C(\mathbf{a}', \mathbf{b}')| \leq 2, \tag{9.34}$$

which is one of the famous Bell inequalities.

It is important to emphasise that the only assumptions that have gone into proving Eq. (9.34) are:

1. For each particle it is meaningful to talk about the actual values of the projection of the spin along any direction.

2. There is locality in the sense that the value of any physical quantity is not changed by altering the position of a remote piece of measuring equipment. This means that both occurrences of a_n in Eq. (9.33) have the same value, *i.e.*, they do not depend on the direction ($\mathbf{b}$ or $\mathbf{b}'$) along which the other observer chooses to measure the spin of particle 2. In particular, we are ruling out the type of context-dependent values that arose in our discussion of the Kochen–Specker theorem.

We shall now show that the predictions of quantum theory violate this inequality over a range of directions for the spin measurements. The quantum-mechanical prediction for the correlation between the spin measurements along axes $\mathbf{a}$ and $\mathbf{b}$ is

$$C(\mathbf{a},\mathbf{b}) := \left(\frac{2}{\hbar}\right)^2 \langle\psi|\, \mathbf{a}\cdot\widehat{\mathbf{S}}_{(1)} \otimes \mathbf{b}\cdot\widehat{\mathbf{S}}_{(2)}\,|\psi\rangle \tag{9.35}$$

where $\widehat{\mathbf{S}}_{(1)}$ and $\widehat{\mathbf{S}}_{(2)}$ are the spin operators for particles 1 and 2 respectively, and the tensor product is defined as in Eq. (8.30). Since the total angular momentum of the vector $|\psi\rangle$ in Eq. (9.29) is zero, it is invariant under the unitary operators which generate rotations of coordinate systems (*cf.* Sections 7.1–7.2). This means that $C(\mathbf{a},\mathbf{b})$ is a function of $\cos\theta_{ab} := \mathbf{a}\cdot\mathbf{b}$ only, and hence there is no loss of generality in assuming that $\mathbf{a}$ points along the z-axis and that $\mathbf{b}$ lies in the x–z plane. Then Eq. (9.35) becomes

$$C(\mathbf{a},\mathbf{b}) = \langle\psi|\, \sigma_{1z} \otimes (\sigma_{2z}\cos\theta_{ab} + \sigma_{2x}\sin\theta_{ab})\,|\psi\rangle \tag{9.36}$$

where, for example, σ_{1z} is the z-direction Pauli spin matrix for the first particle. It is now a straightforward calculation [Exercise] to show that

$$C(\mathbf{a},\mathbf{b}) = -\cos\theta_{ab}. \tag{9.37}$$

Now let us restrict our attention to the special case in which (i) the four vectors $\mathbf{a},\mathbf{a}',\mathbf{b},\mathbf{b}'$ are coplanar; (ii) $\mathbf{a}$ and $\mathbf{b}$ are parallel; and (iii) $\theta_{ab'} = \theta_{a'b} = \phi$ say. Then the Bell inequality will be satisfied provided

$$|1 + 2\cos\phi - \cos 2\phi| \leq 2. \tag{9.38}$$

However, from the form of this function of ϕ sketched in Figure 9.1 we see at once that the inequalities are *violated* for all values of ϕ between 0^0 and 90^0.

This means that if the predictions of quantum theory are experimentally valid in this region then any idea of systems possessing individual values for observables must necessarily involve an essential non-locality. This applies in particular to any hidden variable theory that is completely consistent with the results of quantum theory. Thus the $2^9 5^3$ dollar question is

1. Are the Bell inequalities empirically violated?

2. If so, are such violations in accord with the predictions of quantum theory?

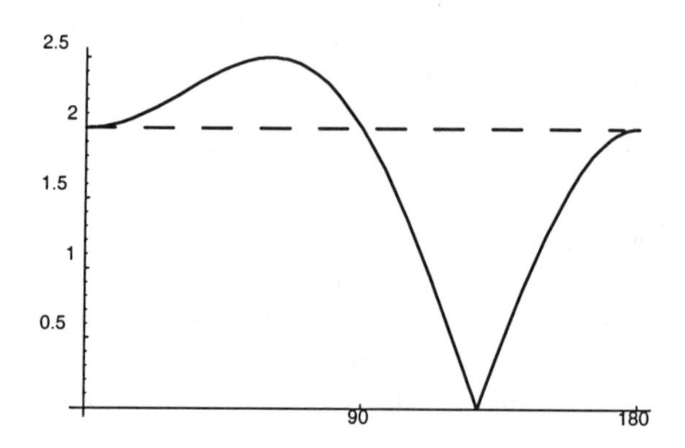

Figure 9.1: The function $|1 + 2\cos\phi - \cos 2\phi|$

A number of experiments have been performed over the last 20 years, the most famous of which is the work of Alain Aspect and collaborators who checked the correlation between photons in a variety of configurations (Aspect, Graingier & Roger 1981), (Aspect, Graingier & Roger 1982*b*), (Aspect, Dalibard & Roger 1982*a*). Photons are much easier to use experimentally than electrons as they pass through air; the analogue of electron spin along a particular axis is played by the polarisation states of the photon. These experiments included ingenious mechanisms whereby the decision of the direction along which to measure the polarisation was deferred until after the photon had left the source, thus ruling out any possibility of a direct, however obscure, causal link between photon production and measurement. In particular, this ensured that the detector settings themselves could not affect the state produced by the decay process.

The details and full implications of these experiments are discussed in several excellent texts, and the interested reader is referred to these (for example, see Redhead (1989), Hughes (1989), and Peres (1993)). In any case, the overwhelming conclusion is that the predictions of quantum theory *are* vindicated, and so we are obliged either to stick to a pragmatic approach or strict instrumentalist interpretation, or else to accept the existence of a strange non-locality that seems hard to reconcile with our normal concepts of spatial separation between independent entities.

9.4 Epilogue

In the course of this book we have discussed many of the central structural features of modern quantum theory; in particular, we have become involved in some of the most important conceptual challenges that confront any attempts to formulate a fully coherent physical interpretation of the extant mathematical formalism. Therefore, it would be nice to finish with a chapter entitled "The Solution to the Conceptual Problems", but unfortunately this is not possible; at least, not in any really comprehensive sense.

Of course, for many physicists, the pragmatic approach to quantum theory is quite acceptable. In the Physics Department at Imperial College we have a large research activity at the frontiers of experimental and theoretical solid-state physics. We also have a major involvement in the experimental programmes at CERN and other international particle-accelerator sites. My colleagues involved in these activities can be, and are, very effective professionally without needing to lose any sleep about the implications of the Kochen–Specher theorem or the Bell inequalities. However, in the last twenty years there has been a growing belief among physicists from many different specialisations that even if modern quantum theory works well at the pragmatic level, it simply cannot be the last word on the matter. The challenge posed by state-reduction, the idea of probability, quantum entanglement, and the role of measurement, cannot be lightly dismissed, and none of the current positions on these matters seems adequate to deserve the accolade of "*The* Solution to the Conceptual Problems".

The central issue in all this is really the phenomenon of quantum entanglement, and its striking contrast with the reductionist concepts of Western philosophy. The intensive study of these matters over the last two decades has clarified considerably the question of what the problems really are. However their resolution is another matter, and there are now many physicists who feel that any significant advance will entail a radical revision in our understanding of the meaning and significance of the categories of space, time and substance. This is a good task for the twenty-first century, and a good thought on which to end these notes.

PROBLEMS & ANSWERS

Problems

1. Show that $\left\{\binom{1}{0}, \binom{i}{1}\right\}$ is a basis set for the vector space $\mathbb{C}^2$, and expand the vectors $\binom{2}{3}$ and $\binom{3i-1}{2i+4}$ in terms of it.

2. Normalise these basis vectors with respect to the usual inner product on $\mathbb{C}^2$.

Is the resulting set orthonormal?

3. Let $\psi(x) = e^{-3x^2}$. What is the 'length' (*i.e.*, norm) of $\vec{\psi}$ in terms of the standard inner product on the space of wave functions?

Does e^{2ix} have a finite length?

Let $\phi(x) := xe^{-x^2}$. What is $\langle\vec{\psi}, \vec{\phi}\rangle$?

(NB. $\int_{-\infty}^{\infty} e^{-ax^2}\, dx = \left(\frac{\pi}{a}\right)^{\frac{1}{2}}$.)

4. Let $\{\vec{e}_1, \vec{e}_2, \ldots, \vec{e}_N\}$ be a basis set for a complex vector space V equipped with a scalar product. Show that if $\vec{\psi} \in V$ is such that, for all $i = 1, 2, \ldots, N$, $\langle\vec{e}_i, \vec{\psi}\rangle = 0$, then $\vec{\psi} = 0$.

5. Show that a scalar product can be defined on the vector space $M(N, \mathbb{C})$ of all $N \times N$ complex matrices by

$$\langle A, B\rangle := \operatorname{tr}(A^\dagger B)$$

where $A, B \in M(N, \mathbb{C})$.

6. Show that any linear operator $\widehat{A}$ on a vector space V equipped with a scalar product is uniquely determined by its matrix elements in the sense that, if $\widehat{A}$ and $\widehat{B}$ are any pair of operators such that $\langle\vec{\psi}, \widehat{A}\vec{\phi}\rangle = \langle\vec{\psi}, \widehat{B}\vec{\phi}\rangle$ for all $\vec{\psi}, \vec{\phi} \in V$, then $\widehat{A} = \widehat{B}$.

7. Show that the components of the matrix representative of an operator $\widehat{A}$ with respect to an orthonormal basis set $\{\vec{e}_1, \vec{e}_2, \ldots, \vec{e}_N\}$ have the explicit form $A_{ij} = \langle\vec{e}_i, \widehat{A}\vec{e}_j\rangle$ where $i, j = 1, 2, \ldots, N$.

8. Show that the adjoints of any pair of operators $\widehat{A}, \widehat{B}$ satisfy the equation $(\widehat{A}\widehat{B})^\dagger = \widehat{B}^\dagger\widehat{A}^\dagger$.

9. The Pauli matrices $\sigma_x, \sigma_y, \sigma_z$ play an important role in quantum theory and are defined in Eq. (1.23). Show that they are self-adjoint operators on $\mathbb{C}^2$, and compute their eigenvalues and eigenvectors.

Check that, for each of the three operators σ_x, σ_y and σ_z, the eigenvectors corresponding to different eigenvalues are orthogonal.

10. What is the explicit form of the most general self-adjoint operator on $\mathbb{C}^2$?

Show how it can be written in terms of the Pauli matrices and the unit 2×2 matrix $\left(\begin{smallmatrix} 1 & 0 \\ 0 & 1 \end{smallmatrix}\right)$.

11. What are the eigenvectors and eigenvalues of the matrix operator $\widehat{M} := \left(\begin{smallmatrix} 0 & 1 \\ 0 & 0 \end{smallmatrix}\right)$?

(i) Are the eigenvectors of $\widehat{M}$ a basis set for $\mathbb{C}^2$?

(ii) Is $\widehat{M}$ a self-adjoint operator?

12. Compute the eigenvectors of the matrix operator

$$\begin{pmatrix} 1 & 0 & 1 \\ 0 & 2 & 0 \\ 1 & 0 & 1 \end{pmatrix}$$

on the vector space $\mathbb{C}^3$.

Construct an orthonormal basis set from the eigenvectors of this operator.

13. Let $\widehat{P}$ be an operator on a Hilbert space $\mathcal{H}$ that satisfies $\widehat{P} = \widehat{P}^{\dagger}$ and $\widehat{P} = \widehat{P}^2$, and which is not the unit operator.

(i) Show that the eigenvalues of $\widehat{P}$ are the numbers 0 and 1.

(ii) Show that $\widehat{P}$ is a projection operator by finding the subspace W of $\mathcal{H}$ onto which it projects.

(iii) For what vectors $|\phi\rangle$ is the operator $|\phi\rangle\langle\phi|$ a projection operator?

14. Let $\widehat{\rho_1}$ and $\widehat{\rho_2}$ be a pair of density matrices. Show that $r\widehat{\rho_1} + (1-r)\widehat{\rho_2}$ is a density matrix for all real numbers r such that $0 \leq r \leq 1$.

15. Show that a density matrix operator $\widehat{\rho}$ represents a vector state (*i.e.*, it can be written as $|\psi\rangle\langle\psi|$ for some vector $|\psi\rangle$) if, and only if, $\widehat{\rho}^2 = \widehat{\rho}$.

16. Let $\widehat{A}$ and $\widehat{B}$ be a pair of self-adjoint operators whose eigenvalues are the discrete sets $\{a_1, a_2, \ldots, a_{M_A}\}$ and $\{b_1, b_2, \ldots, b_{M_B}\}$, and with the associated spectral resolutions $\widehat{A} = \sum_{i=1}^{M_A} a_i \widehat{P}_i$ and $\widehat{B} = \sum_{j=1}^{M_B} b_j \widehat{Q}_j$ respectively.

(i) Show that $[\,\widehat{A}, \widehat{B}\,] = 0$ implies that $[\,\widehat{P}_i, \widehat{B}] =$ for all $i = 1, 2, \ldots, M_A$.

(ii) Hence show that $[\,\widehat{A}, \widehat{B}\,] = 0$ implies that $[\,\widehat{P}_i, \widehat{Q}_j] = 0$ for all $i = 1, 2, \ldots, M_A$ and $j = 1, 2, \ldots, M_B$.

17. Show that the length of a state vector in quantum theory does not change as it evolves in time.

Show that the expected value $\langle A \rangle_{\psi_t}$ of an observable A changes in time in such a way that it satisfies the differential equation

$$i\hbar \frac{d}{dt}\langle A \rangle_{\psi_t} = \langle \vec{\psi}_t, [\,\widehat{A}, \widehat{H}\,]\vec{\psi}_t\rangle$$

where $\widehat{H}$ is the Hamiltonian operator, and the state is normalized to one.

18. If $\widehat{P}$ is a projection operator, and $\widehat{U}$ is any unitary operator, show that $\widehat{U}\widehat{P}\widehat{U}^{-1}$ is also a projection operator.

19. The spin degrees of freedom of a ρ-meson are represented on the vector space $\mathbb{C}^3$ by the matrix operators

$$\widehat{S}_x = \frac{\hbar}{\sqrt{2}}\begin{pmatrix} 0 & 1 & 0 \\ 1 & 0 & 1 \\ 0 & 1 & 0 \end{pmatrix}, \ \widehat{S}_y = \frac{\hbar}{\sqrt{2}}\begin{pmatrix} 0 & -i & 0 \\ i & 0 & -i \\ 0 & i & 0 \end{pmatrix}, \ \widehat{S}_z = \hbar\begin{pmatrix} 1 & 0 & 0 \\ 0 & 0 & 0 \\ 0 & 0 & -1 \end{pmatrix}.$$

Compute the commutator $[\,\widehat{S}_x, \widehat{S}_y\,]$, and check by explicit calculation that the product of the uncertainties in the state

$$|\phi\rangle = \begin{pmatrix} 1 \\ i \\ -2 \end{pmatrix}$$

satisfies the generalized uncertainty relations.

20. For a simple harmonic oscillator (with mass and angular frequency set equal to 1 for convenience) annihilation and creation operators are defined by

$$\widehat{a} := (2\hbar)^{-\frac{1}{2}}(\widehat{x} + i\widehat{p}), \quad \widehat{a}^\dagger := (2\hbar)^{-\frac{1}{2}}(\widehat{x} - i\widehat{p}),$$

and act on the normalised energy eigenvectors $|n\rangle$, $n = 0, 1, \ldots$, as

$$\widehat{a}\,|n\rangle = n^{\frac{1}{2}}\,|n-1\rangle, \quad \widehat{a}^\dagger\,|n\rangle = (n+1)^{\frac{1}{2}}\,|n+1\rangle$$

and with $\widehat{a}\,|0\rangle = 0$.

Use these results to show that the expected values of the observables x and p vanish in these eigenstates, and that the uncertainties satisfy $(\triangle_{|n\rangle}x)(\triangle_{|n\rangle}p) = \frac{\hbar}{2}(2n+1)$.

21. The spin states of an electron are represented by elements of the vector space $\mathbb{C}^2$. Three observables of special interest are the components S_x, S_y and S_z of the spin angular momentum, which are represented by the matrix operators

$$\widehat{S}_x = \frac{\hbar}{2}\sigma_x, \quad \widehat{S}_y = \frac{\hbar}{2}\sigma_y, \quad \widehat{S}_z = \frac{\hbar}{2}\sigma_z$$

where σ_x, σ_y and σ_z are the Pauli spin matrices given in Eq. (1.23).

(i) Suppose that an ensemble of such systems is in a state corresponding to the vector $|\psi\rangle = \binom{1}{0}$. What is the predicted probability that a measurement of S_x, will give the result (a) $\frac{\hbar}{2}$; (b) $-\frac{\hbar}{2}$?

(ii) What would be the analogous probabilities if the observable measured was S_y rather than S_x?

(iii) Suppose that, after making the repeated (ideal) measurements of S_y, the subensemble is selected on which the value $\frac{\hbar}{2}$ has been obtained, and that a measurement of S_z is then made. What is the probability of obtaining the result $\frac{\hbar}{2}$?

22. (a) In the theory of a pair of spinning electrons, compute the following vectors

$$\text{(i)} \quad \widehat{S}_x \otimes \widehat{S}_y \begin{pmatrix} 1 \\ 0 \end{pmatrix} \begin{pmatrix} 0 \\ i \end{pmatrix}$$

$$\text{(ii)} \quad \widehat{\mathbb{1}} \otimes \widehat{S}_z \left\{ \begin{pmatrix} 2 \\ 1 \end{pmatrix} \begin{pmatrix} 0 \\ 3 \end{pmatrix} + 4 \begin{pmatrix} 1 \\ 6 \end{pmatrix} \begin{pmatrix} 1 \\ 1 \end{pmatrix} \right\}$$

$$\text{(iii)} \quad \widehat{S}_x^{\text{tot}} \begin{pmatrix} 1 \\ 1 \end{pmatrix} \begin{pmatrix} 1 \\ 0 \end{pmatrix}$$

where $\widehat{S}_x^{\text{tot}}$ is the *total* spin operator in the x-direction defined by $\widehat{S}_x^{\text{tot}} := \widehat{\mathbb{1}} \otimes \widehat{S}_x + \widehat{S}_x \otimes \widehat{\mathbb{1}}$, and the notation $\binom{a}{b}\binom{c}{d}$ is shorthand for $\binom{a}{b} \otimes \binom{c}{d}$.

(b) Evaluate the following quantities

$$\text{(i)} \quad \left\| \begin{pmatrix} 1 \\ 0 \end{pmatrix} \begin{pmatrix} i \\ 2 \end{pmatrix} \right\|$$

$$\text{(ii)} \quad \left\| \begin{pmatrix} 1 \\ 1 \end{pmatrix} \begin{pmatrix} 1 \\ i \end{pmatrix} \right\|$$

$$\text{(iii)} \quad \left\langle \begin{pmatrix} 1 \\ 0 \end{pmatrix} \begin{pmatrix} i \\ 2 \end{pmatrix}, \begin{pmatrix} 1 \\ 1 \end{pmatrix} \begin{pmatrix} 1 \\ i \end{pmatrix} \right\rangle.$$

Normalise the vector $\binom{1}{0}\binom{i}{2} + \binom{1}{1}\binom{1}{i}$.

(c) If the spin state of an ensemble of pairs of spin-$\frac{1}{2}$ particles is $\binom{1}{0}\binom{i}{2} + \binom{1}{1}\binom{1}{i}$ what are the probabilities that a measurement of S_z^{tot} will yield the values (i) $\hbar$; (ii) $-\hbar$?

Answers

1. If $\binom{1}{0}$ and $\binom{i}{1}$ were linearly dependent there would have to be a complex number μ such that $\binom{1}{0} = \mu\binom{i}{1}$, which is equivalent to the pair of equations $1 = \mu i$ and $0 = \mu$. Clearly this is impossible, and so the two vectors are linearly independent. Hence they form a basis set for the two-dimensional vector space $\mathbb{C}^2$.

(i) Let

$$\begin{pmatrix} 2 \\ 3 \end{pmatrix} = \alpha \begin{pmatrix} 1 \\ 0 \end{pmatrix} + \beta \begin{pmatrix} i \\ 1 \end{pmatrix} \equiv \begin{pmatrix} \alpha + i\beta \\ \beta \end{pmatrix}.$$

Then this is equivalent to the pair of equations $2 = \alpha + i\beta$ and $3 = \beta$, which implies $\alpha = 2 - 3i$. Hence the desired expansion is

$$\begin{pmatrix} 2 \\ 3 \end{pmatrix} = (2 - 3i) \begin{pmatrix} 1 \\ 0 \end{pmatrix} + 3 \begin{pmatrix} i \\ 1 \end{pmatrix}.$$

(ii) Let

$$\begin{pmatrix} 3i - 1 \\ 2i + 4 \end{pmatrix} = \alpha \begin{pmatrix} 1 \\ 0 \end{pmatrix} + \beta \begin{pmatrix} i \\ 1 \end{pmatrix} \equiv \begin{pmatrix} \alpha + i\beta \\ \beta \end{pmatrix}.$$

Thus $3i - 1 = \alpha + i\beta$ and $2i + 4 = \beta$, which implies that $\alpha = 1 - i$. Hence the desired expansion is

$$\begin{pmatrix} 3i - 1 \\ 2i + 4 \end{pmatrix} = (1 - i) \begin{pmatrix} 1 \\ 0 \end{pmatrix} + (2i + 4) \begin{pmatrix} i \\ 1 \end{pmatrix}.$$

2. The norms are

$$\left\| \begin{pmatrix} 1 \\ 0 \end{pmatrix} \right\|^2 = (1,\, 0)^* \begin{pmatrix} 1 \\ 0 \end{pmatrix} = 1,$$

so this vector is already normalised. On the other hand

$$\left\| \begin{pmatrix} i \\ 1 \end{pmatrix} \right\|^2 = (i,\, 1)^* \begin{pmatrix} i \\ 1 \end{pmatrix} = (-i,\, 1) \begin{pmatrix} i \\ 1 \end{pmatrix} = 2,$$

and hence a normalised form of this vector is $\frac{1}{\sqrt{2}}\binom{i}{1}$. (Note that this normalised vector is only defined up to an arbitrary complex phase factor of modulus 1.)

The inner product between the vectors is

$$\left\langle \begin{pmatrix} 1 \\ 0 \end{pmatrix}, \frac{1}{\sqrt{2}} \begin{pmatrix} i \\ 1 \end{pmatrix} \right\rangle = \frac{1}{\sqrt{2}} (1,\, 0)^* \begin{pmatrix} i \\ 1 \end{pmatrix} = \frac{i}{\sqrt{2}}$$

and hence the pair of vectors is *not* orthonormal.

3. (i) If $\psi(x) = e^{-3x^2}$ we have

$$\| \psi \|^2 = \langle \vec{\psi}, \vec{\psi} \rangle = \int_{-\infty}^{\infty} \psi^*(x)\psi(x)\, dx = \int_{-\infty}^{\infty} e^{-6x^2}\, dx = \left(\frac{\pi}{6} \right)^{\frac{1}{2}}$$

and so the length of ψ is

$$\| \psi \| = \left(\frac{\pi}{6} \right)^{\frac{1}{4}}.$$

(ii) If $\phi(x) = e^{2ix}$ we have

$$\| \phi \|^2 = \int_{-\infty}^{\infty} \left(e^{2ix} \right)^* e^{2ix}\, dx = \int_{-\infty}^{\infty} 1\, dx = \infty$$

and hence $\vec{\phi}$ has *infinite* length.

(iii) If $\phi(x) = xe^{-x^2}$ then

$$\langle \vec{\psi}, \vec{\phi} \rangle = \int_{-\infty}^{\infty} \left(e^{-3x^2} \right)^* x\, e^{-x^2}\, dx = \int_{-\infty}^{\infty} x\, e^{-4x^2}\, dx.$$

By inspection, this integral vanishes (since xe^{-4x^2} is an odd function of x), and so

$$\langle \vec{\psi}, \vec{\phi} \rangle = 0.$$

4. Any vector $\vec{\phi} \in V$ can be expanded in the form $\vec{\phi} = \sum_{i=1}^{N} \phi_i \vec{e}_i$ for some unique set of complex numbers $\phi_1, \phi_2, \ldots, \phi_N$. Hence, if $\langle \vec{e}_i, \vec{\psi} \rangle = 0$ for all $i = 1, 2, \ldots, N$, we have

$$\langle \vec{\phi}, \vec{\psi} \rangle = \langle \sum_{i=1}^{N} \phi_i \vec{e}_i, \vec{\psi} \rangle = \sum_{i=1}^{N} \phi_i^* \langle \vec{e}_i, \vec{\psi} \rangle = 0$$

for all $\vec{\phi} \in V$ (I shall not justify the interchange of the scalar product operation with taking the limit $N \mapsto \infty$ in the case when the vector space is infinite-dimensional). Since this is true for all $\vec{\phi} \in V$ it is true in particular for $\vec{\phi} = \vec{\psi}$, and hence $\langle \vec{\psi}, \vec{\psi} \rangle = 0$, which implies $\vec{\psi} = 0$.

5. The vector space structure on $M(N, \mathbb{C})$ is defined by adding the matrix elements of matrices and multiplying them by a scalar. More precisely, if $A, B \in M(N, \mathbb{C})$ and $\lambda \in \mathbb{C}$, we define $A + B$ and λA by

$$(A + B)_{ij} := A_{ij} + B_{ij} \tag{9.1}$$

and

$$(\lambda A)_{ij} := \lambda A_{ij} \tag{9.2}$$

where $i, j = 1, 2, \ldots, N$. To show that $\operatorname{tr}(A^\dagger B)$ is a scalar product we note that $\operatorname{tr} A := \sum_{i=1}^N A_{ii}$ and $(A^\dagger)_{ij} = (A_{ji})^*$. Thus

$$\operatorname{tr}(A^\dagger B) = \sum_{i,j=1}^{N} (A_{ji})^* \, B_{ji},$$

and then the three conditions that must be satisfied by a scalar product are proved as follows.

- It follows from Eq. (9.1–9.2) that if $\alpha, \beta \in \mathbb{C}$, then

  $$\operatorname{tr}\{A^\dagger(\alpha B + \beta C)\} = \sum_{i,j=1}^{N} (A_{ji})^*(\alpha B_{ji} + \beta C_{ji}) = \alpha \operatorname{tr}(A^\dagger B) + \beta \operatorname{tr}(A^\dagger C)$$

  so that $\langle A, (\alpha B + \beta C) \rangle = \alpha \langle A, B \rangle + \beta \langle A, C \rangle$ as required.

- $\left(\operatorname{tr}(A^\dagger B)\right)^* = \left(\sum_{i,j=1}^N (A_{ji})^* B_{ji}\right)^* = \sum_{i,j=1}^N A_{ji}\,(B_{ji})^* = \operatorname{tr}(B^\dagger A)$, so that $\langle A, B \rangle^* = \langle B, A \rangle$ as required.

- Finally, $\operatorname{tr}(A^\dagger A) = \sum_{i,j=1}^N (A_{ji})^* \, A_{ji} = \sum_{i,j=1}^N |A_{ji}|^2$. This is a sum of terms, each of which is positive or zero. Thus $\langle A, A \rangle \geq 0$, and $\langle A, A \rangle = 0$ only if $A_{ji} = 0$ for all $i, j = 1, 2 \ldots, N$.

6. Suppose that $\langle \vec{\psi}, \widehat{A}\vec{\phi} \rangle = \langle \vec{\psi}, \widehat{B}\vec{\phi} \rangle$ for all $\vec{\psi}, \vec{\phi} \in V$. Then, in particular, for all $\vec{\phi} \in V$ we have $\langle \vec{e}_i, \widehat{A}\vec{\phi} \rangle = \langle \vec{e}_i, \widehat{B}\vec{\phi} \rangle$ for all elements $\{\vec{e}_1, \vec{e}_2, \ldots, \vec{e}_N\}$ of a basis set for the vector space V. But then, according to Problem

4, this implies that, for all $\vec{\phi} \in V$, $\hat{A}\vec{\phi} = \hat{B}\vec{\phi}$, which is precisely what is meant by saying that $\hat{A} = \hat{B}$.

7. The matrix elements A_{ij} of an operator $\hat{A}$ with respect to a basis set $\{\vec{e}_1, \vec{e}_2, \ldots, \vec{e}_N\}$ are defined by the equation

$$\hat{A}\vec{e}_j = \sum_{k=1}^{N} \vec{e}_k \, A_{kj}.$$

In the case where the basis set is orthonormal, we can take the scalar product of both sides of this equation with the vector $\vec{e}_i$ to get

$$\begin{aligned}\langle \vec{e}_i, \hat{A}\vec{e}_j \rangle = \langle \vec{e}_i, \sum_{k=1}^{N} \vec{e}_k \, A_{kj} \rangle = \sum_{k=1}^{N} \langle \vec{e}_i, \vec{e}_k \, A_{kj} \rangle = \\ = \sum_{k=1}^{N} \langle \vec{e}_i, \vec{e}_k \rangle A_{kj} = \sum_{k=1}^{N} \delta_{ik} \, A_{kj} = A_{ij}\end{aligned}$$

where we have used the orthonormality condition $\langle \vec{e}_i, \vec{e}_k \rangle = \delta_{ik}$.

8. Let $\vec{\psi}, \vec{\phi}$ be any pair of vectors in the vector space V. Then, by the definition of the adjoint operation, we have

$$\langle \vec{\psi}, (\hat{A}\hat{B})^{\dagger}\vec{\phi} \rangle = \langle \hat{A}\hat{B}\vec{\psi}, \vec{\phi} \rangle = \langle \hat{B}\vec{\psi}, \hat{A}^{\dagger}\vec{\phi} \rangle = \langle \vec{\psi}, \hat{B}^{\dagger}\hat{A}^{\dagger}\vec{\phi} \rangle.$$

Since this is true for all $\vec{\psi}, \vec{\phi} \in V$, the result of Problem 6 shows that $(\hat{A}\hat{B})^{\dagger} = \hat{B}^{\dagger}\hat{A}^{\dagger}$, as required.

9. A self-adjoint operator on $\mathbb{C}^2$ is simply a hermitian 2×2 matrix. That is $A_{ij} = A^*_{ji}$, $i, j = 1, 2$. In other words, if

$$\hat{A} = \begin{pmatrix} \alpha & \beta \\ \gamma & \delta \end{pmatrix} \text{ then } \hat{A}^{\dagger} = \begin{pmatrix} \alpha^* & \gamma^* \\ \beta^* & \delta^* \end{pmatrix}$$

and, if $\hat{A}$ is self-adjoint, then $\hat{A} = \hat{A}^{\dagger}$. Clearly σ_x, σ_y and σ_z all satisfy this condition.

The eigenvalues λ of σ_x are the solutions to

$$\begin{pmatrix} 0 & 1 \\ 1 & 0 \end{pmatrix} \begin{pmatrix} a \\ b \end{pmatrix} = \lambda \begin{pmatrix} a \\ b \end{pmatrix} \text{ that is } \begin{pmatrix} -\lambda & 1 \\ 1 & -\lambda \end{pmatrix} \begin{pmatrix} a \\ b \end{pmatrix} = 0,$$

which is equivalent to $\det\begin{pmatrix} -\lambda & 1 \\ 1 & -\lambda \end{pmatrix} = 0$, that is $\lambda^2 = 1$. Thus the eigenvalues are $\lambda = \pm 1$.

An eigenvector $\begin{pmatrix} a \\ b \end{pmatrix}$ corresponding to $\lambda = 1$ satisfies $\begin{pmatrix} 0 & 1 \\ 1 & 0 \end{pmatrix}\begin{pmatrix} a \\ b \end{pmatrix} = \begin{pmatrix} a \\ b \end{pmatrix}$ which implies that $a = b$. Hence the eigenvectors are of the form $a\begin{pmatrix} 1 \\ 1 \end{pmatrix}$ for any complex number a. A particular one that is normalised is $\frac{1}{\sqrt{2}}\begin{pmatrix} 1 \\ 1 \end{pmatrix}$.

Similarly, a normalised eigenvector for $\lambda = -1$ is $\frac{1}{\sqrt{2}}\begin{pmatrix} 1 \\ -1 \end{pmatrix}$, and

$$\left\langle \frac{1}{\sqrt{2}}\begin{pmatrix} 1 \\ 1 \end{pmatrix}, \frac{1}{\sqrt{2}}\begin{pmatrix} 1 \\ -1 \end{pmatrix} \right\rangle = \frac{1}{2}(1, 1)^* \begin{pmatrix} 1 \\ -1 \end{pmatrix} = 0$$

as expected.

Similarly, the eigenvalues of σ_y are also $\lambda = \pm 1$ with eigenvectors $a\begin{pmatrix} 1 \\ i \end{pmatrix}$ and $b\begin{pmatrix} i \\ 1 \end{pmatrix}$. A normalised set is $\frac{1}{\sqrt{2}}\begin{pmatrix} 1 \\ i \end{pmatrix}$ and $\frac{1}{\sqrt{2}}\begin{pmatrix} i \\ 1 \end{pmatrix}$, and

$$\left\langle \frac{1}{\sqrt{2}}\begin{pmatrix} 1 \\ i \end{pmatrix}, \frac{1}{\sqrt{2}}\begin{pmatrix} i \\ 1 \end{pmatrix} \right\rangle = \frac{1}{2}(1, -i) \begin{pmatrix} i \\ 1 \end{pmatrix} = 0$$

as expected.

Since σ_z is a diagonal matrix its eigenvalues are just its diagonal elements 1 and -1, with corresponding eigenvectors $a\begin{pmatrix} 1 \\ 0 \end{pmatrix}$ and $b\begin{pmatrix} 0 \\ 1 \end{pmatrix}$. A normalised set is $\begin{pmatrix} 1 \\ 0 \end{pmatrix}$ and $\begin{pmatrix} 0 \\ 1 \end{pmatrix}$, and of course

$$\left\langle \begin{pmatrix} 1 \\ 0 \end{pmatrix}, \begin{pmatrix} 0 \\ 1 \end{pmatrix} \right\rangle = (1, 0) \begin{pmatrix} 0 \\ 1 \end{pmatrix} = 0$$

as expected.

10. As shown above, a hermitian 2×2 matrix $\begin{pmatrix} \alpha & \beta \\ \gamma & \delta \end{pmatrix}$ satisfies $\begin{pmatrix} \alpha & \beta \\ \gamma & \delta \end{pmatrix} = \begin{pmatrix} \alpha^* & \gamma^* \\ \beta^* & \delta^* \end{pmatrix}$. Hence the most general self-adjoint 2×2 operator is of the form

$$\begin{pmatrix} r_1 & a \\ a^* & r_2 \end{pmatrix}$$

where a is complex and r_1 and r_2 are real. We can write

$$\begin{pmatrix} r_1 & a \\ a^* & r_2 \end{pmatrix} = \frac{(r_1 + r_2)}{2}\begin{pmatrix} 1 & 0 \\ 0 & 1 \end{pmatrix} + \frac{(r_1 - r_2)}{2}\begin{pmatrix} 1 & 0 \\ 0 & -1 \end{pmatrix} + \\ + \Re a \begin{pmatrix} 0 & 1 \\ 1 & 0 \end{pmatrix} - \Im a \begin{pmatrix} 0 & -i \\ i & 0 \end{pmatrix}$$

where $\Re a$ and $\Im a$ are respectively the real and imaginary parts of the complex number a; that is $a = \Re a + i\Im a$. Hence the desired expression is

$$\begin{pmatrix} r_1 & a \\ a^* & r_2 \end{pmatrix} = \frac{(r_1 + r_2)}{2}\mathbb{1} + \frac{(r_1 - r_2)}{2}\sigma_z + \Re a\,\sigma_x - \Im a\,\sigma_y.$$

11. (i) Let $\begin{pmatrix} 0 & 1 \\ 0 & 0 \end{pmatrix}\begin{pmatrix} a \\ b \end{pmatrix} = \lambda\begin{pmatrix} a \\ b \end{pmatrix}$. Then $\det\begin{pmatrix} -\lambda & 1 \\ 0 & -\lambda \end{pmatrix} = 0$ and so $\lambda^2 = 0$. Hence the only eigenvalue of this operator is 0. This eigenvalue might have been doubly degenerate but, in fact, the equation $\begin{pmatrix} 0 & 1 \\ 0 & 0 \end{pmatrix}\begin{pmatrix} a \\ b \end{pmatrix} = 0$ implies that $b = 0$, and hence the most general eigenvector is $a\begin{pmatrix} 1 \\ 0 \end{pmatrix}$. Thus the eigenvectors do *not* form a basis set.

(ii) The matrix $\begin{pmatrix} 0 & 1 \\ 0 & 0 \end{pmatrix}$ is not of the form $\begin{pmatrix} r_1 & a \\ a^* & r_2 \end{pmatrix}$, and hence it is not hermitian. Thus there is no contradiction with the basic expansion theorem.

12. The eigenvalues of the matrix $\begin{pmatrix} 1 & 0 & 1 \\ 0 & 2 & 0 \\ 1 & 0 & 1 \end{pmatrix}$ are the roots of the algebraic equation

$$\det\begin{pmatrix} 1-\lambda & 0 & 1 \\ 0 & 2-\lambda & 0 \\ 1 & 0 & 1-\lambda \end{pmatrix} = 0$$

which is $(1-\lambda)^2(2-\lambda) - (2-\lambda) = 0$, that is $\lambda(\lambda-2)(2-\lambda) = 0$. Thus the eigenvalues are 0 and 2, with the second one appearing twice. For $\lambda = 0$, an eigenvector obeys

$$\begin{pmatrix} 1 & 0 & 1 \\ 0 & 2 & 0 \\ 1 & 0 & 1 \end{pmatrix}\begin{pmatrix} a \\ b \\ c \end{pmatrix} = 0$$

which gives $a + c = 0$ and $2b = 0$. Thus the eigenvectors are of the form $a\begin{pmatrix} 1 \\ 0 \\ -1 \end{pmatrix}$, with a normalised example being

$$\frac{1}{\sqrt{2}}\begin{pmatrix} 1 \\ 0 \\ -1 \end{pmatrix}.$$

For $\lambda = 2$, the eigenvector equation is

$$\begin{pmatrix} 1 & 0 & 1 \\ 0 & 2 & 0 \\ 1 & 0 & 1 \end{pmatrix} \begin{pmatrix} a \\ b \\ c \end{pmatrix} = 2 \begin{pmatrix} a \\ b \\ c \end{pmatrix}$$

which implies that $a + c = 2a$ with b undetermined. Hence the eigenvectors are of the form $\begin{pmatrix} a \\ b \\ a \end{pmatrix}$, and any two such, $\begin{pmatrix} a \\ b \\ a \end{pmatrix}$ and $\begin{pmatrix} a' \\ b' \\ a' \end{pmatrix}$, will be orthogonal if

$$(a^*, b^*, a^*) \begin{pmatrix} a' \\ b' \\ a' \end{pmatrix} = 0$$

which implies that $2a^*a' + b^*b' = 0$. Two linearly independent (and normalised) such vectors are

$$\frac{1}{\sqrt{3}} \begin{pmatrix} 1 \\ 1 \\ 1 \end{pmatrix} \text{ and } \frac{1}{\sqrt{6}} \begin{pmatrix} 1 \\ -2 \\ 1 \end{pmatrix}.$$

Thus the eigenvalue $\lambda = 2$ is doubly degenerate, and the three vectors constructed above form an orthonormal basis set for $\mathbb{C}^3$.

13. (i) Suppose that $\hat{P}\vec{u} = \mu\vec{u}$ so that μ is an eigenvalue of $\hat{P}$ with eigenvector $\vec{u}$. Then

$$\mu\vec{u} = \hat{P}\vec{u} = \hat{P}^2\vec{u} = \hat{P}(\hat{P}\vec{u}) = \hat{P}(\mu\vec{u}) = \mu^2\vec{u}$$

and hence, since $\vec{u} \neq 0$, w~ have $\mu - \mu^2 = 0$, that is $\mu(1 - \mu) = 0$. Thus the two possible eigenvalues for $\hat{P}$ are $\mu = 1$ and $\mu = 0$.

Both do actually occur if $P \neq \hat{\mathbb{1}}$ since, if $\vec{\psi}$ is any vector such that $\hat{P}\vec{\psi} \neq 0$ (and there must be at least one, or else $\hat{P}$ would be the null operator) then $\hat{P}(\hat{P}\vec{\psi}) = \hat{P}^2\vec{\psi} = \hat{P}\vec{\psi}$ so that $\hat{P}\vec{\psi}$ is an eigenvector of $\hat{P}$ with eigenvalue $\mu = 1$.

Now choose any $\vec{\psi}$ such that $\hat{P}\vec{\psi} \neq \vec{\psi}$ (and there must be at least one, or else $\hat{P}$ would be the unit operator). Then $\hat{P}(\hat{\mathbb{1}} - \hat{P})\vec{\psi} = (\hat{P} - \hat{P}^2)\vec{\psi} = 0$, and hence the non-null vector $(\hat{\mathbb{1}} - \hat{P})\vec{\psi}$ is an eigenvector of $\hat{P}$ with eigenvalue 0. Thus the eigenvalues of $\hat{P}$ are 0 and 1.

(ii) Let us define the subspace $W := \{\vec{\psi} \in \mathcal{H} | \widehat{P}\vec{\psi} = \vec{\psi}\}$, *i.e.*, the set of all eigenvectors of $\widehat{P}$ with eigenvalue 1. Any vector $\vec{\psi} \in \mathcal{H}$ can be written as the identity

$$\vec{\psi} \equiv \widehat{P}\vec{\psi} + (\widehat{\mathbb{1}} - \widehat{P})\vec{\psi} \tag{9.3}$$

and, since $\widehat{P}^2 = \widehat{P}$, we see that $\widehat{P}\vec{\psi}$ belongs to W. Furthermore, since $\widehat{P}$ is self-adjoint, for any $\vec{\psi}, \vec{\phi} \in \mathcal{H}$ we have

$$\langle \widehat{P}\vec{\psi}, (\widehat{\mathbb{1}} - \widehat{P})\vec{\phi} \rangle = \langle (\widehat{\mathbb{1}} - \widehat{P})\widehat{P}\vec{\psi}, \vec{\phi} \rangle = 0$$

where we have used the result $(\widehat{\mathbb{1}} - \widehat{P})\widehat{P} = \widehat{P} - \widehat{P}^2 = \widehat{P} - \widehat{P} = 0$. Thus Eq. (9.3) gives a decomposition of the vector $\vec{\psi}$ into a part $\widehat{P}\vec{\psi}$ in W, and a part $(\widehat{\mathbb{1}} - \widehat{P})\vec{\psi}$ in the orthogonal complement $W^\perp$ of W. However, it is easy to show that, given any subspace W, a decomposition of any vector $\vec{\psi} = \vec{\psi}_W + \vec{\psi}_{W^\perp}$ into parts $\vec{\psi}_W \in W$ and $\vec{\psi}_{W^\perp} \in W^\perp$ is unique. Thus Eq. (9.3) is the unique such decomposition, and so $\vec{\psi}_W = P\vec{\psi}$ for all vectors $\vec{\psi}$. Hence $\widehat{P}$ is the operator that projects onto the subspace W. Thus $\widehat{P}$ is indeed a projection operator.

(iii) The adjoint of the general operator $|\psi\rangle\langle\phi|$ is $|\phi\rangle\langle\psi|$, and hence $|\phi\rangle\langle\phi|$ is self-adjoint for any vector $|\phi\rangle$. Furthermore, $(\,|\phi\rangle\langle\phi|\,)^2 = (\langle\phi|\phi\rangle)\,|\phi\rangle\langle\phi|$, and hence $(\,|\phi\rangle\langle\phi|\,)^2 = |\phi\rangle\langle\phi|$ if and only if $|\phi\rangle$ is normalised; that is $\langle\phi|\phi\rangle = 1$. Therefore, this is the condition for $|\phi\rangle\langle\phi|$ to be a projection operator.

14. Let $\widehat{\rho} := r\widehat{\rho_1} + (1-r)\widehat{\rho_2}$ where $r \in \mathbb{R}$ is such that $0 \leq r \leq 1$. To be a density matrix, $\widehat{\rho}$ must satisfy the three conditions

- $\widehat{\rho} = \widehat{\rho}^\dagger$;
- $\widehat{\rho}$ is a positive, semi-definite operator;
- $\operatorname{tr}\widehat{\rho} = 1$.

The first condition follows for any real r since $(r\widehat{\rho_1} + (1-r)\widehat{\rho_2})^\dagger = r\widehat{\rho_1}^\dagger + (1-r)\widehat{\rho_2}^\dagger = r\widehat{\rho_1} + (1-r)\widehat{\rho_2}$ because $\widehat{\rho_1}$ and $\widehat{\rho}_2$ are self-adjoint.

To show that the second condition is satisfied, let $|\psi\rangle$ be any non-null vector in the Hilbert space. Then, since $\widehat{\rho_1}$ and $\widehat{\rho_2}$ are positive semi-definite operators, we have $\langle\psi|\widehat{\rho_1}|\psi\rangle \geq 0$ and $\langle\psi|\widehat{\rho_2}|\psi\rangle \geq 0$. Furthermore, if $0 \leq r \leq 1$, then $0 \leq (1-r) \leq 1$. Thus

$$\langle\psi|\widehat{\rho}|\psi\rangle = r\langle\psi|\widehat{\rho_1}|\psi\rangle + (1-r)\langle\psi|\widehat{\rho_2}|\psi\rangle$$

is a sum of two numbers, both of which are ≥ 0. Hence the sum is itself ≥ 0.

To show that the third condition is satisfied we use the fact that the trace is a linear operation. Thus

$$\mathrm{tr}\,(r\widehat{\rho_1} + (1-r)\widehat{\rho_2}) = r\,\mathrm{tr}\,\widehat{\rho_1} + (1-r)\mathrm{tr}\,\widehat{\rho_2} = (1-r) + r = 1$$

where we have also used the condition $\mathrm{tr}\,\widehat{\rho_1} = 1 = \mathrm{tr}\,\widehat{\rho_2}$ on the two density matrices.

15. First, assume that $\widehat{\rho}$ represents a vector state, *i.e.*, $\widehat{\rho} = \widehat{P}_{|\psi\rangle} := |\psi\rangle\langle\psi|$ for some normalised state $|\psi\rangle$. Then

$$(\widehat{\rho})^2 = |\psi\rangle\langle\psi|\;|\psi\rangle\langle\psi| = |\psi\rangle\langle\psi| = \widehat{\rho}$$

using the normalisation condition $\langle\psi|\psi\rangle = 1$.

Conversely, suppose that the density matrix $\widehat{\rho}$ is such that $(\widehat{\rho})^2 = \widehat{\rho}$. We know that $\widehat{\rho}$ can be written in the form $\widehat{\rho} = \sum_{d=1}^{D} w_d\,\widehat{P}_{|\psi_d\rangle}$ for some collection of normalised states $\{\,|\psi_1\rangle, |\psi_2\rangle, \ldots, |\psi_D\rangle\}$ and real numbers $\{w_1, w_2, \ldots, w_D\}$ with $0 < w_d \leq 1$ for $d = 1, 2, \ldots, D$. Now, since the collection of states $\{\,|\psi_1\rangle, |\psi_2\rangle, \ldots, |\psi_D\rangle\}$ is orthonormal, we have $\widehat{P}_{|\psi_c\rangle}\,\widehat{P}_{|\psi_d\rangle} = \delta_{cd}\widehat{P}_{|\psi_d\rangle}$, and hence

$$(\widehat{\rho})^2 = \sum_{c=1}^{D}\sum_{d=1}^{D} w_c w_d\,\widehat{P}_{|\psi_c\rangle}\widehat{P}_{|\psi_d\rangle} = \sum_{d=1}^{D}(w_d)^2\,\widehat{P}_{|\psi_d\rangle}.$$

However, since $(\widehat{\rho})^2 = \widehat{\rho}$, this expression is also equal to $\sum_{d=1}^{D} w_d\widehat{P}_{|\psi_d\rangle}$. Thus

$$\sum_{d=1}^{D}(w_d)^2\,\widehat{P}_{|\psi_d\rangle} = \sum_{d=1}^{D} w_d\,\widehat{P}_{|\psi_d\rangle},$$

and then multiplying both sides on the left by $\widehat{P}_{|\psi_c\rangle}$ for $c = 1, 2, \ldots, D$, and using $\widehat{P}_{|\psi_c\rangle}\,\widehat{P}_{|\psi_d\rangle} = \delta_{cd}\widehat{P}_{\psi_d}$, we find that $((w_c)^2 - w_c)\widehat{P}_{|\psi_c\rangle} = 0$ for all $c = 1, 2, \ldots D$, which implies that, for all c, $w_c^2 = w_c$. The only solution to this equation is $w_c = 1$ or $w_c = 0$, which, since $w_c > 0$, implies that $w_c = 1$. However, $\sum_{d=1}^{D} w_d = 1$, and so $D = 1$. Therefore $\widehat{\rho} = \widehat{P}_{|\psi\rangle}$ for some vector $|\psi\rangle$, *i.e.*, $\widehat{\rho}$ represents a vector state.

16. (i) We have $\widehat{A} = \sum_{i=1}^{M_A} a_i \widehat{P}_i$ and $\widehat{B} = \sum_{j=1}^{M_B} b_j \widehat{Q}_j$ with $[\widehat{A}, \widehat{B}] = 0$. Thus, for all $\vec{\psi} \in \mathcal{H}$ and $i = 1, 2, \ldots, M_A$ we have $\widehat{A}(\widehat{B}\widehat{P}_i\vec{\psi}) = \widehat{B}\widehat{A}\widehat{P}_i\vec{\psi} = a_i(\widehat{B}\widehat{P}_i\vec{\psi})$ (since $\widehat{A}\widehat{P}_i = a_i\widehat{P}_i$), and so $\widehat{B}\widehat{P}_i\vec{\psi}$ belongs to the eigenspace of $\widehat{A}$ associated with the eigenvalue a_i. Hence,

$$\widehat{P}_i\widehat{B}\widehat{P}_i\vec{\psi} = \widehat{B}\widehat{P}_i\vec{\psi} \tag{9.4}$$

which, since it is true for all $\vec{\psi} \in \mathcal{H}$, means that

$$\widehat{P}_i\widehat{B}\widehat{P}_i = \widehat{B}\widehat{P}_i. \tag{9.5}$$

Taking the adjoint of this equation gives $\widehat{P}_i\widehat{B}\widehat{P}_i = \widehat{P}_i\widehat{B}$, and then subtracting these two equations gives $\widehat{B}\widehat{P}_i = \widehat{P}_i\widehat{B}$, as required.

(ii) We can now apply the same argument starting with the equation $[\widehat{B}, \widehat{P}_i] = 0$ and using the spectral projectors $\widehat{Q}_j$ of $\widehat{B}$. The result is that $[\widehat{Q}_j, \widehat{P}_i] = 0$, as required.

17. The length squared of a vector $\vec{\psi}_t$ is $\| \psi_t \|^2$, and the vector itself satisfies the time-dependent Schrödinger equation $i\hbar\frac{d}{dt}\vec{\psi}_t = \widehat{H}\vec{\psi}_t$ where $\widehat{H}$ is the Hamiltonian operator. Then

$$\begin{aligned}
\frac{d}{dt} \| \psi_t \|^2 &= \frac{d}{dt}\langle\vec{\psi}_t, \vec{\psi}_t\rangle = \langle\frac{d}{dt}\vec{\psi}_t, \vec{\psi}_t\rangle + \langle\vec{\psi}_t, \frac{d}{dt}\vec{\psi}_t\rangle \\
&= \langle-\frac{i}{\hbar}\widehat{H}\vec{\psi}_t, \vec{\psi}_t\rangle + \langle\vec{\psi}_t, -\frac{i}{\hbar}\widehat{H}\vec{\psi}_t\rangle \\
&= \frac{i}{\hbar}\left(\langle\widehat{H}\vec{\psi}_t, \vec{\psi}_t\rangle - \langle\vec{\psi}_t, \widehat{H}\vec{\psi}_t\rangle\right) = 0
\end{aligned}$$

where the last equality follows from the fact that $\widehat{H}$ is self-adjoint. Thus the length of the vector $\vec{\psi}_t$ is constant in time.

Alternatively, the result follows immediately from the fact that $\vec{\psi}_t = e^{-it\widehat{H}/\hbar}\vec{\psi}_0$, and $e^{-it\widehat{H}/\hbar}$ is a unitary operator.

The proof of the result for the expected value is similar to the above. Thus, if the state is normalised at time $t = 0$, the result above shows that it is normalised for all time, and so $\langle A \rangle_{\psi_t} = \langle \vec{\psi}_t, \widehat{A}\vec{\psi}_t \rangle$. Then

$$\begin{aligned} i\hbar \frac{d}{dt}\langle A \rangle_{\psi_t} &= i\hbar \left(\langle \frac{d}{dt}\vec{\psi}_t, \widehat{A}\vec{\psi}_t \rangle + \langle \vec{\psi}_t, \widehat{A}\frac{d}{dt}\vec{\psi}_t \rangle \right) \\ &= \langle -\widehat{H}\vec{\psi}_t, \widehat{A}\vec{\psi}_t \rangle + \langle \vec{\psi}_t, \widehat{A}\widehat{H}\vec{\psi}_t \rangle \\ &= -\langle \vec{\psi}_t, \widehat{H}\widehat{A}\vec{\psi}_t \rangle + \langle \vec{\psi}_t, \widehat{A}\widehat{H}\vec{\psi}_t \rangle \\ &= \langle \vec{\psi}_t, [\,\widehat{A}, \widehat{H}\,]\vec{\psi}_t \rangle, \end{aligned}$$

which is the desired result.

18. An operator $\widehat{P}$ is a projection operator if, and only if, $\widehat{P} = \widehat{P}^\dagger$ and $(\widehat{P})^2 = \widehat{P}$. Let $\widehat{Q} := \widehat{U}\widehat{P}\widehat{U}^{-1}$. Then, since $\widehat{U}$ is unitary, we have $\widehat{U}^{-1} = \widehat{U}^\dagger$ and hence $\widehat{Q} = \widehat{U}\widehat{P}\widehat{U}^\dagger$. Furthermore, in general we have $(\widehat{A}\widehat{B})^\dagger = \widehat{B}^\dagger\widehat{A}^\dagger$, and hence

$$\widehat{Q}^\dagger = (\widehat{U}\widehat{P}\widehat{U}^\dagger)^\dagger = (\widehat{U}^\dagger)^\dagger\widehat{P}^\dagger\widehat{U}^\dagger = \widehat{U}\widehat{P}\widehat{U}^\dagger$$

where we have used the identity $\widehat{A} = (\widehat{A}^\dagger)^\dagger$ which is true for any operator $\widehat{A}$. Thus $\widehat{Q} = \widehat{Q}^\dagger$, *i.e.*, $\widehat{U}\widehat{P}\widehat{U}^{-1}$ is self-adjoint.

Furthermore

$$(\widehat{U}\widehat{P}\widehat{U}^{-1})^2 = \widehat{U}\widehat{P}\widehat{U}^{-1}\widehat{U}\widehat{P}\widehat{U}^{-1} = \widehat{U}\widehat{P}^2\widehat{U}^{-1} = \widehat{U}\widehat{P}\widehat{U}^{-1},$$

so that $\widehat{U}\widehat{P}\widehat{U}^{-1}$ is indeed a projection operator.

19. The desired commutator is $\widehat{S}_x\widehat{S}_y - \widehat{S}_y\widehat{S}_x$, and direct calculation gives

$$\widehat{S}_x\widehat{S}_y = \frac{\hbar^2}{2}\begin{pmatrix} 0 & 1 & 0 \\ 1 & 0 & 1 \\ 0 & 1 & 0 \end{pmatrix}\begin{pmatrix} 0 & -i & 0 \\ i & 0 & -i \\ 0 & i & 0 \end{pmatrix} = \frac{\hbar^2}{2}\begin{pmatrix} i & 0 & -i \\ 0 & 0 & 0 \\ i & 0 & -i \end{pmatrix}$$

and

$$\widehat{S}_y\widehat{S}_x = \frac{\hbar^2}{2}\begin{pmatrix} 0 & -i & 0 \\ i & 0 & -i \\ 0 & i & 0 \end{pmatrix}\begin{pmatrix} 0 & 1 & 0 \\ 1 & 0 & 1 \\ 0 & 1 & 0 \end{pmatrix} = \frac{\hbar^2}{2}\begin{pmatrix} -i & 0 & -i \\ 0 & 0 & 0 \\ i & 0 & i \end{pmatrix}.$$

Thus

$$[\,\widehat{S}_x, \widehat{S}_y\,] = \frac{\hbar^2}{2}\begin{pmatrix} 2i & 0 & 0 \\ 0 & 0 & 0 \\ 0 & 0 & -2i \end{pmatrix} = i\hbar^2\begin{pmatrix} 1 & 0 & 0 \\ 0 & 0 & 0 \\ 0 & 0 & -1 \end{pmatrix} = i\hbar\widehat{S}_z.$$

The general form of the uncertainty relation is

$$\Delta_\psi S_x \, \Delta_\psi S_y \;\geq\; \frac{1}{2}|\langle\psi|\,[\,\widehat{S}_x, \widehat{S}_y\,]\,|\psi\rangle| = \frac{\hbar}{2}|\langle\psi|\,\widehat{S}_z\,|\psi\rangle|$$

and $(\Delta_\psi S_x)^2 = \langle S_x^2\rangle_\psi - \langle S_x\rangle_\psi^2$. Therefore the first step is to calculate $\langle S_x\rangle_\psi$ in the normalised state

$$|\psi\rangle = \frac{1}{\sqrt{6}}\begin{pmatrix} 1 \\ i \\ -2 \end{pmatrix}.$$

This is

$$\begin{aligned}
\langle S_x\rangle_\psi = \langle\psi|\,\widehat{S}_x\,|\psi\rangle &= \frac{\hbar}{6\sqrt{2}}(1,\,-i,\,-2)\begin{pmatrix} 0 & 1 & 0 \\ 1 & 0 & 1 \\ 0 & 1 & 0 \end{pmatrix}\begin{pmatrix} 1 \\ i \\ -2 \end{pmatrix} \\
&= \frac{\hbar}{6\sqrt{2}}(1,\,-i,\,-2)\begin{pmatrix} i \\ -1 \\ i \end{pmatrix} = 0
\end{aligned}$$

and hence $(\Delta_\psi S_x)^2 = \langle S_x^2\rangle_\psi$. Now,

$$\widehat{S}_x^2 = \frac{\hbar^2}{2}\begin{pmatrix} 0 & 1 & 0 \\ 1 & 0 & 1 \\ 0 & 1 & 0 \end{pmatrix}\begin{pmatrix} 0 & 1 & 0 \\ 1 & 0 & 1 \\ 0 & 1 & 0 \end{pmatrix} = \frac{\hbar^2}{2}\begin{pmatrix} 1 & 0 & 1 \\ 0 & 2 & 0 \\ 1 & 0 & 1 \end{pmatrix}$$

and so

$$\begin{aligned}
\langle S_x^2\rangle_\psi &= \frac{\hbar^2}{12}(1,\,-i,\,-2)\begin{pmatrix} 1 & 0 & 1 \\ 0 & 2 & 0 \\ 1 & 0 & 1 \end{pmatrix}\begin{pmatrix} 1 \\ i \\ -2 \end{pmatrix} \\
&= \frac{\hbar^2}{12}(1,\,-i,\,-2)\begin{pmatrix} -1 \\ 2i \\ -1 \end{pmatrix} = \frac{\hbar^2}{4}
\end{aligned}$$

so that

$$\Delta_\psi S_x = \frac{\hbar}{2}.$$

Now we must compute $(\Delta_\psi S_y)^2 = \langle S_y^2\rangle_\psi - \langle S_y\rangle_\psi^2$. We have

$$\langle S_y\rangle_\psi = \langle\psi|\,\widehat{S}_x\,|\psi\rangle = \frac{\hbar}{6\sqrt{2}}(1,\,-i,\,-2)\begin{pmatrix} 0 & -i & 0 \\ i & 0 & -i \\ 0 & i & 0 \end{pmatrix}\begin{pmatrix} 1 \\ i \\ -2 \end{pmatrix}$$

$$= \frac{\hbar}{6\sqrt{2}} (1, -i, -2) \begin{pmatrix} 1 \\ 3i \\ -1 \end{pmatrix} = \frac{\hbar}{\sqrt{2}}$$

and

$$\widehat{S}_y^2 = \frac{\hbar^2}{2} \begin{pmatrix} 0 & -i & 0 \\ i & 0 & -i \\ 0 & i & 0 \end{pmatrix} \begin{pmatrix} 0 & -i & 0 \\ i & 0 & -i \\ 0 & i & 0 \end{pmatrix} = \frac{\hbar^2}{2} \begin{pmatrix} 1 & 0 & -1 \\ 0 & 2 & 0 \\ -1 & 0 & 1 \end{pmatrix}$$

so that

$$\begin{aligned} \langle S_y^2 \rangle_\psi &= \frac{\hbar^2}{12} (1, -i, -2) \begin{pmatrix} 1 & 0 & -1 \\ 0 & 2 & 0 \\ -1 & 0 & 1 \end{pmatrix} \begin{pmatrix} 1 \\ i \\ -2 \end{pmatrix} \\ &= \frac{\hbar^2}{12} (1, -i, -2) \begin{pmatrix} 3 \\ 2i \\ -3 \end{pmatrix} = \frac{\hbar^2}{12} 11. \end{aligned}$$

Hence

$$(\Delta_\psi S_y)^2 = \frac{11}{12}\hbar^2 - \left(\frac{\hbar}{\sqrt{2}}\right)^2 = \frac{5}{12}\hbar^2$$

so that

$$\Delta_\psi S_y = \hbar \left(\frac{5}{12}\right)^{\frac{1}{2}}.$$

Hence the left hand side of the uncertainty relation is $\Delta_\psi S_x \, \Delta_\psi S_y = \frac{\hbar^2}{2}\left(\frac{5}{12}\right)^{\frac{1}{2}} = \hbar^2 \left(\frac{5}{48}\right)^{\frac{1}{2}}$. On the other hand, the right hand side is

$$\begin{aligned} \frac{\hbar}{2} |\langle \psi | \, \widehat{S}_z \, | \psi \rangle| &= \frac{\hbar}{12} \left| (1, -i - 2) \begin{pmatrix} 1 & 0 & 0 \\ 0 & 0 & 0 \\ 0 & 0 & -1 \end{pmatrix} \begin{pmatrix} 1 \\ i \\ -2 \end{pmatrix} \right| \\ &= \frac{\hbar}{12} \left| (1, -i - 2) \begin{pmatrix} 1 \\ 0 \\ 2 \end{pmatrix} \right| = \frac{\hbar^2}{4}. \end{aligned}$$

But $\frac{\hbar^2}{4} = \hbar^2 \left(\frac{4}{48}\right)^{\frac{1}{2}} < \hbar^2 \left(\frac{5}{48}\right)^{\frac{1}{2}}$, and hence the uncertainty relations are satisfied.

20. We have $\widehat{a} := (2\hbar)^{-\frac{1}{2}}(\widehat{x} + i\widehat{p})$ and $\widehat{a}^\dagger := (2\hbar)^{-\frac{1}{2}}(\widehat{x} - i\widehat{p})$, so that $\widehat{x} = (\hbar/2)^{\frac{1}{2}}(\widehat{a} + \widehat{a}^\dagger)$ and $\widehat{p} = (\hbar/2)^{\frac{1}{2}}(\widehat{a} - \widehat{a}^\dagger)/i$. Then

$$\widehat{x}\,|n\rangle = (\hbar/2)^{\frac{1}{2}}\left\{n^{\frac{1}{2}}\,|n-1\rangle + (n+1)^{\frac{1}{2}}\,|n+1\rangle\right\}$$

and so $\langle n|\,\widehat{x}\,|n\rangle = (\hbar/2)^{\frac{1}{2}}\left\{n^{\frac{1}{2}}\langle n|n-1\rangle + (n+1)^{\frac{1}{2}}\langle n|n+1\rangle\right\} = 0$, where I have used $\langle n|m\rangle = \delta_{nm}$ for the eigenvectors of a self-adjoint operator (the Hamiltonian) corresponding to different eigenvalues. Thus the expected value of x is 0.

Similarly, $\widehat{p}\,|n\rangle = (\hbar/2)^{\frac{1}{2}}\frac{1}{i}\left\{n^{\frac{1}{2}}\,|n-1\rangle - (n+1)^{\frac{1}{2}}\,|n+1\rangle\right\}$, and hence, for the same reason as above, $\langle n|\,\widehat{p}\,|n\rangle = 0$ for all energy eigenstates $|n\rangle$. Thus the expected value of p is also 0 for all states $|n\rangle$.

To compute the uncertainties, we first use the above expressions for x to give

$$\begin{aligned}(\triangle_n x)^2 &= \langle n|\,\widehat{x}^2\,|n\rangle - (\langle n|\,\widehat{x}\,|n\rangle)^2 = \langle n|\,\widehat{x}^2\,|n\rangle \\ &= \frac{\hbar}{2}\langle n|\,(\widehat{a} + \widehat{a}^\dagger)^2\,|n\rangle = \langle n|\,\widehat{a}^2 + (\widehat{a}^\dagger)^2 + \widehat{a}\widehat{a}^\dagger + \widehat{a}^\dagger\widehat{a}\,|n\rangle.\end{aligned}$$

But $\widehat{a}^2\,|n\rangle = n^{\frac{1}{2}}(n-1)^{\frac{1}{2}}\,|n-2\rangle$ and $(\widehat{a}^\dagger)^2\,|n\rangle = (n+1)^{\frac{1}{2}}(n+2)^{\frac{1}{2}}\,|n+2\rangle$, and hence the first two terms vanish (because $\langle n|n-2\rangle = 0 = \langle n|n+2\rangle$). Thus

$$(\triangle_n x)^2 = \frac{\hbar}{2}\langle n|\,(n+1)^{\frac{1}{2}}(n+1)^{\frac{1}{2}} + n^{\frac{1}{2}}n^{\frac{1}{2}}\,|n\rangle = \frac{\hbar}{2}(2n+1).$$

Similarly,

$$(\triangle_n p)^2 = -\frac{\hbar}{2}\langle n|\,(\widehat{a} - \widehat{a}^\dagger)(\widehat{a} - \widehat{a}^\dagger)\,|n\rangle = \frac{\hbar}{2}\langle n|\,(\widehat{a}\widehat{a}^\dagger + \widehat{a}^\dagger\widehat{a})\,|n\rangle = \frac{\hbar}{2}(2n+1),$$

and hence $(\triangle_n x)(\triangle_n p) = \frac{\hbar}{2}(2n+1)$.

21. (i) From Problem 9 we know that $\widehat{S}_x = \frac{\hbar}{2}\begin{pmatrix}0&1\\1&0\end{pmatrix}$ has eigenvalues $\hbar/2$ and $-\hbar/2$, with normalised eigenvectors $\frac{1}{\sqrt{2}}\begin{pmatrix}1\\1\end{pmatrix}$ and $\frac{1}{\sqrt{2}}\begin{pmatrix}1\\-1\end{pmatrix}$ respectively. Then, according to the general formalism, if the state of an ensemble is $|\psi\rangle = \begin{pmatrix}1\\0\end{pmatrix}$,

$$\mathrm{Prob}(S_x = \frac{\hbar}{2};\begin{pmatrix}1\\0\end{pmatrix}) = \left|\left\langle \frac{1}{\sqrt{2}}\begin{pmatrix}1\\1\end{pmatrix}, \begin{pmatrix}1\\0\end{pmatrix}\right\rangle\right|^2 = \frac{1}{2}$$

and

$$\text{Prob}(S_x = \frac{-\hbar}{2}; \begin{pmatrix}1\\0\end{pmatrix}) = \left|\left\langle \frac{1}{\sqrt{2}}\begin{pmatrix}1\\-1\end{pmatrix}, \begin{pmatrix}1\\0\end{pmatrix}\right\rangle\right|^2 = \frac{1}{2}.$$

(Note that these probabilities add up to 1. This is correct since $\hbar/2$ and $-\hbar/2$ are the only two possible values for the result of the measurement of S_x.)

(ii) If S_y is measured then the vectors of interest are the eigenvectors of $\widehat{S}_y = \frac{\hbar}{2}\begin{pmatrix}0 & -i\\ i & 0\end{pmatrix}$. These are $\frac{1}{\sqrt{2}}\begin{pmatrix}1\\i\end{pmatrix}$ and $\frac{1}{\sqrt{2}}\begin{pmatrix}i\\1\end{pmatrix}$, corresponding to eigenvalues $\hbar/2$ and $-\hbar/2$ respectively. Thus

$$\text{Prob}(S_y = \frac{\hbar}{2}; \begin{pmatrix}1\\0\end{pmatrix}) = \left|\left\langle \frac{1}{\sqrt{2}}\begin{pmatrix}1\\i\end{pmatrix}, \begin{pmatrix}1\\0\end{pmatrix}\right\rangle\right|^2 = \frac{1}{2}$$

and

$$\text{Prob}(S_y = \frac{-\hbar}{2}; \begin{pmatrix}1\\0\end{pmatrix}) = \left|\left\langle \frac{1}{\sqrt{2}}\begin{pmatrix}i\\1\end{pmatrix}, \begin{pmatrix}1\\0\end{pmatrix}\right\rangle\right|^2 = \frac{1}{2}.$$

(iii) If a measurement of S_z is made following the ideal measurement of S_y, the state of the subensemble for predicting the results of the former is $\frac{1}{\sqrt{2}}\begin{pmatrix}1\\i\end{pmatrix}$ (a 'collapse' of the state vector from $\begin{pmatrix}1\\0\end{pmatrix}$ to the eigenvector of S_y with eigenvalue $+\hbar/2$). Thus

$$\text{Prob}(S_z = \frac{\hbar}{2}; \frac{1}{\sqrt{2}}\begin{pmatrix}1\\i\end{pmatrix}) = \left|\left\langle \begin{pmatrix}1\\0\end{pmatrix}, \frac{1}{\sqrt{2}}\begin{pmatrix}1\\i\end{pmatrix}\right\rangle\right|^2 = \frac{1}{2}.$$

Note that this is less than 1, whereas the state $\begin{pmatrix}1\\0\end{pmatrix}$ with which we started is an eigenvector of $\widehat{S}_z$ with eigenvalue $\hbar/2$, and hence would have yielded $\hbar/2$ with probability 1 as the result of a measurement.

22. (a) (i) The definition of the tensor product $\widehat{A} \otimes \widehat{B}$ of two operators $\widehat{A}$ and $\widehat{B}$ is $\widehat{A} \otimes \widehat{B}(\vec{\phi} \otimes \vec{\psi}) := (\widehat{A}\vec{\phi}) \otimes (\widehat{B}\vec{\psi})$. Hence

$$\widehat{S}_x \otimes \widehat{S}_y \begin{pmatrix}1\\0\end{pmatrix}\begin{pmatrix}0\\i\end{pmatrix} = \frac{\hbar^2}{4}\left[\begin{pmatrix}0 & 1\\1 & 0\end{pmatrix}\begin{pmatrix}1\\0\end{pmatrix}\right]\left[\begin{pmatrix}0 & -i\\i & 0\end{pmatrix}\begin{pmatrix}0\\i\end{pmatrix}\right] = \frac{\hbar^2}{4}\begin{pmatrix}0\\1\end{pmatrix}\begin{pmatrix}1\\0\end{pmatrix}.$$

(ii) Similarly,

$$\begin{aligned}
&\hat{\mathbb{1}} \otimes \hat{S}_z \left\{ \begin{pmatrix} 2 \\ 1 \end{pmatrix} \begin{pmatrix} 0 \\ 3 \end{pmatrix} + 4 \begin{pmatrix} 1 \\ 6 \end{pmatrix} \begin{pmatrix} 1 \\ 1 \end{pmatrix} \right\} \\
&= \frac{\hbar}{2} \left\{ \begin{pmatrix} 2 \\ 1 \end{pmatrix} \left[\begin{pmatrix} 1 & 0 \\ 0 & -1 \end{pmatrix} \begin{pmatrix} 0 \\ 3 \end{pmatrix} \right] + 4 \begin{pmatrix} 1 \\ 6 \end{pmatrix} \left[\begin{pmatrix} 1 & 0 \\ 0 & -1 \end{pmatrix} \begin{pmatrix} 1 \\ 1 \end{pmatrix} \right] \right\} \\
&= \frac{\hbar}{2} \left\{ \begin{pmatrix} 2 \\ 1 \end{pmatrix} \begin{pmatrix} 0 \\ -3 \end{pmatrix} + 4 \begin{pmatrix} 1 \\ 6 \end{pmatrix} \begin{pmatrix} 1 \\ -1 \end{pmatrix} \right\}.
\end{aligned}$$

(iii) We have $\hat{S}_x^{\text{tot}} = \hat{\mathbb{1}} \otimes \hat{S}_x + \hat{S}_x \otimes \hat{\mathbb{1}}$ and hence

$$\begin{aligned}
\hat{S}_x^{\text{tot}} \begin{pmatrix} 1 \\ 1 \end{pmatrix} \begin{pmatrix} 1 \\ 0 \end{pmatrix} &= \frac{\hbar}{2} \left\{ \begin{pmatrix} 1 \\ 1 \end{pmatrix} \left[\begin{pmatrix} 0 & 1 \\ 1 & 0 \end{pmatrix} \begin{pmatrix} 1 \\ 0 \end{pmatrix} \right] + \left[\begin{pmatrix} 0 & 1 \\ 1 & 0 \end{pmatrix} \begin{pmatrix} 1 \\ 1 \end{pmatrix} \right] \begin{pmatrix} 1 \\ 0 \end{pmatrix} \right\} \\
&= \frac{\hbar}{2} \left\{ \begin{pmatrix} 1 \\ 1 \end{pmatrix} \begin{pmatrix} 0 \\ 1 \end{pmatrix} + \begin{pmatrix} 1 \\ 1 \end{pmatrix} \begin{pmatrix} 1 \\ 0 \end{pmatrix} \right\} = \frac{\hbar}{2} \begin{pmatrix} 1 \\ 1 \end{pmatrix} \begin{pmatrix} 1 \\ 1 \end{pmatrix}.
\end{aligned}$$

(b) (i) We have

$$\begin{aligned}
\left\| \begin{pmatrix} 1 \\ 0 \end{pmatrix} \begin{pmatrix} i \\ 2 \end{pmatrix} \right\|^2 &= \left\langle \begin{pmatrix} 1 \\ 0 \end{pmatrix} \begin{pmatrix} i \\ 2 \end{pmatrix}, \begin{pmatrix} 1 \\ 0 \end{pmatrix} \begin{pmatrix} i \\ 2 \end{pmatrix} \right\rangle \\
&= \left\langle \begin{pmatrix} 1 \\ 0 \end{pmatrix}, \begin{pmatrix} 1 \\ 0 \end{pmatrix} \right\rangle \left\langle \begin{pmatrix} i \\ 2 \end{pmatrix}, \begin{pmatrix} i \\ 2 \end{pmatrix} \right\rangle = 1 \times 5 = 5
\end{aligned}$$

and so

$$\left\| \begin{pmatrix} 1 \\ 0 \end{pmatrix} \begin{pmatrix} i \\ 2 \end{pmatrix} \right\| = \sqrt{5}.$$

(ii) Similarly

$$\left\| \begin{pmatrix} 1 \\ 1 \end{pmatrix} \begin{pmatrix} 1 \\ i \end{pmatrix} \right\|^2 = \left\| \begin{pmatrix} 1 \\ 1 \end{pmatrix} \right\|^2 \left\| \begin{pmatrix} 1 \\ i \end{pmatrix} \right\|^2 = 2 \times 2 = 4$$

and so

$$\left\| \begin{pmatrix} 1 \\ 1 \end{pmatrix} \begin{pmatrix} 1 \\ i \end{pmatrix} \right\| = 2.$$

(iii) The inner product is

$$\left\langle \begin{pmatrix} 1 \\ 0 \end{pmatrix} \begin{pmatrix} i \\ 2 \end{pmatrix}, \begin{pmatrix} 1 \\ 1 \end{pmatrix} \begin{pmatrix} 1 \\ i \end{pmatrix} \right\rangle = \left\langle \begin{pmatrix} 1 \\ 0 \end{pmatrix}, \begin{pmatrix} 1 \\ 1 \end{pmatrix} \right\rangle \left\langle \begin{pmatrix} i \\ 2 \end{pmatrix}, \begin{pmatrix} 1 \\ i \end{pmatrix} \right\rangle$$

which is simply

$$
\begin{aligned}
(1,0)^* \begin{pmatrix}1\\1\end{pmatrix} \times (i,2)^* \begin{pmatrix}1\\i\end{pmatrix} &= (1,0) \begin{pmatrix}1\\1\end{pmatrix} \times (-i,2) \begin{pmatrix}1\\i\end{pmatrix} \\
&= 1 \times (-i + 2i) = i.
\end{aligned}
$$

Let $\vec{\psi} := \vec{u} + \vec{v}$ where $\vec{u} := \binom{1}{0}\binom{i}{2}$ and $\vec{v} := \binom{1}{1}\binom{1}{i}$. Then

$$
\begin{aligned}
\| \psi \|^2 &= \langle \vec{\psi}, \vec{\psi} \rangle = \langle \vec{u} + \vec{v}, \vec{u} + \vec{v} \rangle \\
&= \| u \|^2 + \| v \|^2 + \langle \vec{u}, \vec{v} \rangle + \langle \vec{v}, \vec{u} \rangle \\
&= \| u \|^2 + \| v \|^2 + \langle \vec{u}, \vec{v} \rangle + \langle \vec{u}, \vec{v} \rangle^* \\
&= 5 + 4 + i - i = 9.
\end{aligned}
$$

Hence a normalised version of $\vec{\psi}$ is

$$
\vec{\psi} := \frac{1}{3}\left\{ \begin{pmatrix}1\\0\end{pmatrix}\begin{pmatrix}i\\2\end{pmatrix} + \begin{pmatrix}1\\1\end{pmatrix}\begin{pmatrix}1\\i\end{pmatrix} \right\}.
$$

(c) The eigenvectors of $\widehat{S}_z^{\text{tot}} := \widehat{\mathbb{1}} \otimes \widehat{S}_z + \widehat{S}_z \otimes \widehat{\mathbb{1}}$ are clearly $|\uparrow\uparrow\rangle := \binom{1}{0}\binom{1}{0}$ and $|\downarrow\downarrow\rangle := \binom{0}{1}\binom{0}{1}$ with eigenvalues $\hbar$ and $-\hbar$ respectively. The normalised state is $|\psi\rangle := \frac{1}{3}\left\{\binom{1}{0}\binom{i}{2} + \binom{1}{1}\binom{1}{i}\right\}$. Then, according to the general rules of quantum theory

$$
\begin{aligned}
&\text{Prob}(S_z = \hbar;\ |\psi\rangle) = |\langle \uparrow\uparrow |\psi\rangle|^2 \\
&= \frac{1}{9} \left| \left\langle \begin{pmatrix}1\\0\end{pmatrix}\begin{pmatrix}1\\0\end{pmatrix}, \begin{pmatrix}1\\0\end{pmatrix}\begin{pmatrix}i\\2\end{pmatrix} + \begin{pmatrix}1\\1\end{pmatrix}\begin{pmatrix}1\\i\end{pmatrix} \right\rangle \right|^2 \\
&= \frac{1}{9} \left| \left\langle \begin{pmatrix}1\\0\end{pmatrix}\begin{pmatrix}1\\0\end{pmatrix}, \begin{pmatrix}1\\0\end{pmatrix}\begin{pmatrix}i\\2\end{pmatrix} \right\rangle + \left\langle \begin{pmatrix}1\\0\end{pmatrix}\begin{pmatrix}1\\0\end{pmatrix}, \begin{pmatrix}1\\1\end{pmatrix}\begin{pmatrix}1\\i\end{pmatrix} \right\rangle \right|^2 \\
&= \frac{1}{9} \left| \left\langle \begin{pmatrix}1\\0\end{pmatrix}, \begin{pmatrix}1\\0\end{pmatrix} \right\rangle \left\langle \begin{pmatrix}1\\0\end{pmatrix}, \begin{pmatrix}i\\2\end{pmatrix} \right\rangle + \left\langle \begin{pmatrix}1\\0\end{pmatrix}, \begin{pmatrix}1\\1\end{pmatrix} \right\rangle \left\langle \begin{pmatrix}1\\0\end{pmatrix}, \begin{pmatrix}1\\i\end{pmatrix} \right\rangle \right|^2 \\
&= \frac{1}{9}|i + 1|^2 = \frac{2}{9}.
\end{aligned}
$$

Similarly,

$$
\text{Prob}(S_z = -\hbar;\ |\psi\rangle) = |\langle \downarrow\downarrow |\psi\rangle|^2
$$

$$
\begin{aligned}
&= \frac{1}{9}\left|\left\langle \begin{pmatrix}0\\1\end{pmatrix}\begin{pmatrix}0\\1\end{pmatrix}, \begin{pmatrix}1\\0\end{pmatrix}\begin{pmatrix}i\\2\end{pmatrix} + \begin{pmatrix}1\\1\end{pmatrix}\begin{pmatrix}1\\i\end{pmatrix}\right\rangle\right|^2 \\
&= \frac{1}{9}\left|\left\langle \begin{pmatrix}0\\1\end{pmatrix}, \begin{pmatrix}1\\0\end{pmatrix}\right\rangle \left\langle \begin{pmatrix}0\\1\end{pmatrix}, \begin{pmatrix}i\\2\end{pmatrix}\right\rangle + \left\langle \begin{pmatrix}0\\1\end{pmatrix}, \begin{pmatrix}1\\1\end{pmatrix}\right\rangle \left\langle \begin{pmatrix}0\\1\end{pmatrix}, \begin{pmatrix}1\\i\end{pmatrix}\right\rangle\right|^2 \\
&= \frac{1}{9}|0+i|^2 = \frac{1}{9}.
\end{aligned}
$$

Bibliography

Aspect, A., Dalibard, J. & Roger, G. (1982*a*), 'Experimental tests of Bell's inequalities using time-varying analysers', *Phys. Rev. Lett.* **49**, 1804–1807.

Aspect, A., Graingier, P. & Roger, G. (1981), 'Experimental tests of realistic local theories via Bell's theorem', *Phys. Rev. Lett.* **47**, 460–467.

Aspect, A., Graingier, P. & Roger, G. (1982*b*), 'Experimental realization of EPR Gedankenexperiment: A new violation of Bell's inequalities', *Phys. Rev. Lett.* **49**, 91–94.

Bell, J. (1987), *Speakable and Unspeakable in Quantum Mechanics*, Cambridge University Press, Cambridge.

Beltrametti, E. & Cassinelli, G. (1981), *The Logic of Quantum Mechanics*, Addison-Wesley, London.

Birkhoff, G. & von Neumann, J. (1936), 'The logic of quantum mechanics', *Annals of Mathematics* **37**, 823–843.

Bohm, D. (1951), *Quantum Theory*, Prentice-Hall, New Jersey.

Bohm, D. & Hiley, B. (1993), *The Undivided Universe: An Ontological Interpretation of Quantum theory*, Routledge, London.

Borges, J. (1964), *Dreamtigers*, University of Texas Press, Austin, Texas. Transl. by M. Boyer & H. Morland.

Busch, P., Lahti, P. & Mittelstaedt, P. (1991), *The Quantum Theory of Measurement*, Springer-Verlag, London.

DeWitt, B. & Graham, N. (1973), *The Many-Worlds Interpretation of Quantum Mechanics*, Princeton University Press, Princeton.

Dirac, P. (1958), *The Principles of Quantum Mechanics*, Oxford University Press, Oxford.

Eddington, A. (1920), *Space, Time and Gravitation*, Cambridge University Press, Cambridge. p201.

Einstein, A., Podolsky, B. & Rosen, N. (1935), 'Can quantum-mechanical description of reality be considered complete?', *Phys. Rev.* **47**, 777–780.

Everett, H. (1957), 'Relative state formulation of quantum mechanics', *Rev. Mod. Phys.* **29**, 141–149.

Fine, A. (1986), *The Shaky Game: Einstein, Realism and the Quantum Theory*, University of Chicago Press, Chicago.

Ghirardi, G., Rimini, A. & Weber, T. (1986), 'Unified dynamics for microscopic and macroscopic systems', *Phys. Rev.* **D34**, 470–491.

Gleason, A. (1957), 'Measures on the closed subspaces of a Hilbert space', *Journal of Mathematics and Mechanics* **6**, 885–893.

Grayling, A. (1990), *An Introduction to Philosophical Logic*, Duckworth, London.

Hartle, J. (1968), 'Quantum mechanics of individual systems', *Am. J. Phys.* **36**, 704–712.

Heidegger, M. (1967), *What is a Thing?*, Regnery/Gateway Press, Indiana. Transl. W.B. Barton & V. Deutsch. First published in German as *Die Frage nach dem Ding*.

Heisenberg, W. (1952), *Philosophic Problems of Nuclear Science*, Pantheon, New York.

Holland, P. (1993), *The Quantum Theory of Motion*, Cambridge University Press, Cambridge.

Hughes, R. (1989), *The Structure and Interpretation of Quantum Mechanics*, Harvard University Press, Harvard.

Isham, C. (1989), *Lectures on Groups and Vector Spaces*, World Scientific, London.

Isham, C. (1994), Prima facie questions in quantum gravity, *in* J. Ehlers & H. Friedrich, eds, 'Canonical Relativity: Classical and Quantum', Springer-Verlag, Berlin, pp. 1–21.

Jauch, J. (1973), *Foundations of Quantum Mechanics*, Addison-Wesley, London.

Joos, E. & Zeh, H. (1985), 'The emergence of classical properties through interaction with the environment', *Zeitschrift für Physik* **B59**, 223–243.

Jung, C. (1983), Thoughts on the nature and value of speculative enquiry, *in* 'The Zofingia Lectures', Routledge & Kegan Paul, London.

Kochen, S. & Specher, E. (1967), 'The problem of hidden variables in quantum mechanics', *Journal of Mathematics and Mechanics* **17**, 59–87.

Lockwood, M. (1992), *Mind, Brain and the Quantum: The Compound I*, Blackwell, Oxford.

Lüders, G. (1951), 'Über die zustandsänderung durch den messprozess', *Annalen der Physik.* **8**, 322–328.

Margenau, H. (1949), 'Reality in quantum mechanics', *Phil. Science* **16**, 287–302.

Margenau, H. (1963*a*), 'Measurements and quantum states. Part I', *Phil. Science* **30**, 1–16.

Margenau, H. (1963*b*), 'Measurements and quantum states. Part II', *Phil. Science* **30**, 138–157.

Margenau, H. (1963*c*), 'Measurements in quantum mechanics', *Ann. Phys.* **23**, 469–485.

McKnight, J. (1952), 'An extended latency interpretation of quantum mechanical measurement', *Phil. Science* **25**, 209–222.

Omnès, R. (1992), 'Consistent interpretations of quantum mechanics', *Rev. Mod. Phys.* **64**, 339–382.

Omnès, R. (1994), *The Interpretation of Quantum Mechanics*, Princeton University Press, Princeton.

Park, J. (1968*a*), 'Quantum theoretical concepts of measurement: Part I', *Phil. Science* **35**, 205–231.

Park, J. (1968*b*), 'Quantum theoretical concepts of measurement: Part II', *Phil. Science* **35**, 389–411.

Park, J. & Margenau, H. (1968), 'Simultaneous measurement in quantum theory', *Int. J. Theor. Phys.* **1**, 211–283.

Pearle, P. (1986), Models for reduction, *in* R. Penrose & C. Isham, eds, 'Quantum Concepts in Space and Time', Clarendon Press, Oxford, pp. 84–108.

Penrose, R. (1989), *The Emperor's New Mind*, Oxford University Press, Oxford.

Peres, A. (1991), 'Two simple proofs of the Kochen-Specker theorem', *Journal of Physics A* **24**, L175–L178.

Peres, A. (1993), *Quantum Theory: Concepts and Methods*, Kluwer Academic, Boston.

Popper, S. (1956), *Quantum Theory and the Schism in Physics*, Hutchinson, London.

Putnam, H. (1957), 'Three-valued logic', *Phil. Studies* **VIII**, 73–80.

Redhead (1989), *Incompleteness, Nonlocality, and Realism*, Clarendon Press, Oxford.

Reed, M. & Simon, B. (1972), *Methods of Mathematical Physics. I: Functional Analysis*, Academic Press, New York.

Reichenbach, H. (1944), *Philosophical Foundations of Quantum Mechanics*, University of California Press, Berkeley, California.

Scheibe, E. (1973), *The Logical Analysis of Quantum Mechanics*, Pergamon Press, Oxford.

Stapp, H. (1993), *Mind, Matter, and Quantum Mechanics*, Springer-Verlag, London.

Van Frassen, B. (1991), *Quantum Mechanics: An Empiricist View*, Clarendon Press, Oxford.

von Neumann, J. (1971), *Mathematical Foundations of Quantum Mechanics*, Princeton University Press, Princeton.

Wheeler, J. & Zurek, W. (1983), *Quantum Theory and Measurement*, Princeton University Press, Princeton.

Wigner, E. (1961), Remarks on the mind-body question, *in* I. Good, ed., 'The Scientist Speculates', W. Heinemann, London.

Wigner, E. (1963), 'The problem of measurement', *American Journal of Physics* **31**, 6–15.

Williams, C. (1968), *Many Dimensions*, Faber, London.

Index